EMC for Product Designers

EMC for Product Designers

Tim Williams

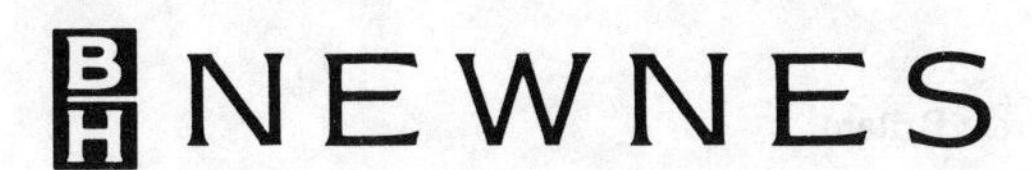

Newnes
An imprint of Butterworth-Heinemann Ltd
Linacre House, Jordan Hill, Oxford OX2 8DP

 PART OF REED INTERNATIONAL BOOKS

OXFORD LONDON BOSTON
MUNICH NEW DELHI SINGAPORE SYDNEY
TOKYO TORONTO WELLINGTON

First published 1992

British Library Cataloguing in Publication Data
A catalogue record for this book is available from the British Library

Library of Congress Cataloguing in Publication Data
A catalogue record for this book is available from the Library of Congress

ISBN 0 7506 1264 9

Printed and bound in Great Britain.

Contents

Preface

The most popular TV programme in the UK over Christmas 1991 was *Auntie's Bloomers*. Millions of viewers watched a selection of clips from the BBC's archives, showing various well known television personalities at embarrassing moments while the camera was running, clips that never made it to the final programme. For a significant fraction of these viewers, their enjoyment would have been spoilt by a bloomer of another kind. In the first half of 1985, before a service charge was introduced, the UK's Radio Investigation Service was receiving an average of 1900 complaints a month relating to broadcast reception, and about 80% of the RIS's resources were devoted to domestic radio and TV reception problems. Spots, hash, snowstorms, colour and vision distortion and occasionally complete loss of picture are all symptoms of the same cause – electromagnetic interference.

It is irritating for the viewer when the picture flickers or is wiped out during a crucial programme, just as it is irritating for a music lover who has carefully taped an important broadcast on FM radio only to find that the quiet passages are ruined by the intrusion of the neighbour's electric drill. It is far more critical when the emergency services are unable to communicate within a city centre because their radio signals are obscured by the electromagnetic "smog" emitted by thousands of computer terminals in the buildings around them.

The coexistence of all kinds of radio services, which use the electromagnetic spectrum to convey information, with technical processes and products from which electromagnetic energy is an undesirable by-product, creates the problem of what is known as *electromagnetic compatibility* (EMC). The solution is a compromise: radio services must allow for a certain degree of interference, but interfering emissions may not exceed a certain level, which normally involves measures to limit or suppress the interference energy. There is an economic tradeoff inherent in this compromise. A lower level of interference would mean that less powerful transmitters were necessary, but the suppression costs would be higher. Alternatively, accept high power transmitters – with the attendant inefficient spectrum usage – in return for lower suppression costs. This economic balance has been tested over the past decades with the establishment of various standards for allowable levels of interference.

The problems of EMC are not limited to interference with radio services. Increasingly, electronic equipment of all kinds is becoming more susceptible to malfunctions caused by external interference. This phenomenon is more and more noticeable for two reasons: the greater pervasiveness and interaction of electronic products in all aspects of daily life, and the relatively worse immunity of modern equipment using plastic cases and microprocessors. Susceptibility to interference is now an issue for many kinds of electronic device, especially those whose continued correct operation is vital for safety or economic reasons. Automotive and aviation control systems are examples of the former category, banking and telecommunication networks are examples of the latter.

There is now an urgent need for mandatory measures to be taken to protect and ensure equipment's EMC. Various national administrations have taken *ad hoc* measures in the past to impose restrictions on some of the electromagnetic properties of some types of product. These measures have often come to be seen as implementing back-door methods of protectionism, with the technical inadequacy of some of the requirements allowing effectively different standards to be applied to imported and indigenous products.

In an heroic effort to recognise the need for EMC protection measures and at the same time to eliminate the protectionist barriers to trade throughout the European Community, the European Commission adopted in 1989 a Directive "on the approximation of the laws of the Member States relative to electromagnetic compatibility", otherwise known as the EMC Directive. It is discussed in detail in chapter 1 of this book.

The adjective "heroic" is used above because the eventual implementation of the requirements of the EMC Directive is proving to be a task worthy of any Hercules. Both the scope of the Directive and the EMC phenomena it covers are exhaustive, but it is framed in extremely general terms. The interpretation of these terms is taking considerable effort, as is the generation of standards against which compliance with the Directive can be judged. As an example of the latter, CENELEC (the European standards body) was mandated to produce several new standards within two years, when the normal process for generating international standards takes at least five years.

Practically speaking, the new standards will start appearing in a steady stream during 1992 and succeeding years, and will undoubtedly be subject to continual revision as experience is gained with their use. One consequence of the lack of availability of these standards is that the Directive's initial timetable, to be fully in place by the end of 1992, has been extended to include a transitional period of three or four years during which its observance is optional.

The task facing manufacturers who must comply with the Directive is, many feel, equally heroic. There is virtually no type of product for which the Directive's requirements are being met in their entirety already. Many manufacturers of data processing or household equipment already meet emission standards required by American, German or EC legislation and for them it is, in the words of one commentator, "business as usual". But the Directive also requires equal attention to be given to a product's immunity. Few products which meet emission standards have also been tested for immunity. There are some product sectors which already exercise immunity standards on a contractual basis – but few of these also test for emissions. This is only pragmatic, since emissions legislation concentrates on protecting the innocent spectrum user while immunity standards are intended to safeguard the product's own end user, and it is still at present rare for these needs to co-exist in the same environment. For the first time, the Directive brings together mandatory requirements for both emission suppression and immunity.

By 1995, every company that manufactures or imports electrical or electronic products will need to have in place measures that will enable its products to comply with the Directive. This means that an awareness of EMC will have to penetrate every part of the enterprise. EMC is undoubtedly affected by the design of the product, and the design and development group is where the awareness normally starts. But it also depends on the way an individual product is put together, so it affects the production department; by the way it is installed, so it affects the installation and service technicians, and the user documentation; it needs to be assured for each unit, so it affects the test department; it impacts the product's marketing strategy and sales literature, so it affects the sales and marketing departments; and it ultimately affects the viability and liabilities of the company, so it must be understood by the senior management.

There are various means of implanting and cultivating this awareness. The many EMC training courses and awareness seminars are a good starting point. It would be possible to bring in consultants to handle every aspect of the EMC compliance process, but for many products this would be expensive and cumbersome and would not necessarily result in improved awareness and expertise within the company where it was really needed. It would also be possible to send every appropriate member of staff on a training course. This would certainly raise awareness but it may not prove so effective in the long run, since EMC techniques also need to be practised to be properly understood. It would also be expensive, but some of the larger companies with established in house training programmes are capable of taking this route.

A good compromise is to nominate one person, or a group if the resources are available, to act as the centre of EMC expertise for the company. His, her or its responsibility should be to implement the requirements of the EMC Directive and any other EMC specifications to which the company may need to work. In the long term, it should also be to make the EMC centre redundant: to imbue a knowledge of EMC principles into each operating division so that they are a natural part of the functioning of that division. This, though, would take years of continuous oversight and education. Meanwhile, the tasks would include:

- reviewing each new product design throughout the development and prototyping stages for adherence to EMC principles, and advising on design changes where necessary;
- drawing up and implementing an EMC test and control plan for each product;
- supervising pre-compliance and compliance tests both in house and in liaison with external test houses;
- maintaining an intimate knowledge of the EMC standards and legislation that apply to the company's products;
- liaising with marketing, sales, production, test, installation and servicing departments to ensure that their strategies are consistent with EMC requirements.

There are probably more detailed tasks involved, but this serves as an indication of the breadth of scope of the EMC engineer's job. It is comparable to that of the quality department, and indeed can sometimes be incorporated within that department.

This book is intended to help the work of the company's EMC centre. It can be used as a reference for the EMC engineer, as background reading for designers and technicians new to the subject, or as part of the armoury of the development group tackling a new project. It is structured into two parts. The first part (Chapters 1–3) discusses the European legislative framework now being erected to encompass EMC, and the test techniques that are used to demonstrate compliance with that framework. It is mainly non-technical in nature. Chapter 1 introduces the subject of interference, and goes on to discuss the provisions of the EMC Directive and the means of achieving compliance with it. Chapter 2 details the various standards that are now in existence or at an advanced stage of drafting, which are relevant for compliance with the Directive. A brief comparison with American and German standards is also included. Chapter 3 covers the test methods for RF emissions, RF susceptibility, ESD and transient susceptibility and mains input current harmonics that are laid down in the standards and which will need to be followed both in house and by external test houses.

The second part of the book discusses techniques for achieving an acceptable EMC performance at minimum extra cost, at the design stage. It is usually possible to add screening and suppression components to an existing design to enable it to meet EMC standards. This brute force method is expensive, time consuming and inefficient. Far better is to design to the appropriate principles from the start, so that the product has a good chance of achieving compliance first time, or if it doesn't then modifications are made easy to implement.

Chapter 4 covers the basic principles involved in coupling electromagnetic interference from a source to a victim. Chapter 5 looks at the techniques which can be applied before resorting to the more traditional methods of screening and suppression: attention to layout and grounding, and choice of circuit configuration, components and software features. Chapter 6 carries on to discuss the accepted "special" EMC techniques which include cable configuration and termination, filtering methods and components, and shielding.

Finally, a series of appendices gather together some of the more detailed reference information.

Many aspects of the EMC Directive are still fluid at the time of writing, such as the actual legislation which will implement it, the detailed contents of the technical construction file and the nature of the penalties for non compliance. Much of the detail of implementation will only become clear when it is actually operating, and cases of non compliance are tested legally. At the same time, the standards will undergo further revision in the light of experience, and new standards will be published. Therefore this book must only be viewed as a guide, and cannot be regarded as a definitive reference in these areas.

Much of the book grew out of course notes that were prepared for seminars on Design for EMC, and I am grateful to those designers who attended these seminars and stimulated me to continually improve and hone the presentation. Many people helped with queries in particular areas during the preparation of the book, but I must particularly acknowledge and thank Ray Hughes, Graham Mays and David Riley of Chase EMC for their support and encouragement. Finally, love and thanks to Joy, without whom many things would not have happened.

Tim Williams
March 1992

Part 1

The Directive, Standards and Testing

Chapter 1

Introduction

1.1 What is EMC?

Electromagnetic interference (EMI) is a serious and increasing form of environmental pollution. Its effects range from minor annoyances due to crackles on broadcast reception, to potentially fatal accidents due to corruption of safety-critical control systems. Various forms of EMI may cause electrical and electronic malfunctions, can prevent the proper use of the radio frequency spectrum, can ignite flammable or other hazardous atmospheres, and may even have a direct effect on human tissue. As electronic systems penetrate more deeply into all aspects of society, so both the potential for interference effects and the potential for serious EMI-induced incidents will increase.

Some reported examples of electromagnetic incompatibility are:

- in Germany, a particular make of car would stall on a stretch of Autobahn opposite a high power broadcast transmitter. Eventually that section of the motorway had to be screened with wire mesh;
- on another type of car, the central door locking and electric sunroof would operate when the car's mobile transmitter was used;
- new electronic push-button telephones installed near the Brookmans Park medium wave transmitter in North London were constantly afflicted with BBC radio programmes;
- mobile phones have been found to interfere with the readings of certain types of petrol pump meter;
- in America, police departments complained that coin-operated electronic games were causing harmful interference to their highway communications system;
- interference to aeronautical safety communications at a US airport was traced to an electronic cash register a mile away;
- the instrument panel of a well known airliner was said to carry the warning "ignore all instruments while transmitting HF";
- electronic point-of-sale units used in shoe, clothing and optician shops (where thick carpets and nylon-coated assistants were common) would experience lock up, false data and uncontrolled drawer openings;
- when a piezo-electric cigarette lighter was lit near the cabinet of a car park barrier control box, the radiated pulse caused the barrier to open and drivers were able to park free of charge;
- lowering the pantographs of electric locomotives at British Rail's Liverpool

Street station interfered with newly installed signalling control equipment, causing the signals to "fail safe" to red;

- perhaps the most tragic example was the fate of HMS Sheffield in the Falklands war, when the missile warning radar that could have detected the Exocet missile which sank the ship was turned off because it interfered with the ship's satellite communications system.

1.1.1 Compatibility between systems

The threat of EMI is controlled by adopting the practices of electromagnetic *compatibility* (EMC). This is defined [98] as "the ability of a device, unit of equipment or system to function satisfactorily in its electromagnetic environment without introducing intolerable electromagnetic disturbances to anything in that environment". The term EMC has two complementary aspects:

- it describes the ability of electrical and electronic systems to operate without interfering with other systems;
- it also describes the ability of such systems to operate as intended within a specified electromagnetic environment.

Thus it is closely related to the environment within which the system operates. Effective EMC requires that the system is designed, manufactured and tested with regard to its predicted operational electromagnetic environment: that is, the totality of electromagnetic phenomena existing at its location. Although the term "electromagnetic" tends to suggest an emphasis on high frequency field-related phenomena, in practice the definition of EMC encompasses all frequencies and coupling paths, from DC to 400GHz.

1.1.1.1 Subsystems within an installation

There are two approaches to EMC. In one case the nature of the installation determines the approach. EMC is especially problematic when several electronic or electrical systems are packed in to a very compact installation, such as on board aircraft, ships, satellites or other vehicles. In these cases susceptible systems may be located very close to powerful emitters and special precautions are needed to maintain compatibility. To do this cost-effectively calls for a detailed knowledge of both the installation circumstances and the characteristics of the emitters and their potential victims. Military, aerospace and vehicle EMC specifications have evolved to meet this need and are well established in their particular industry sectors.

Since this book is concerned with product design to meet the EMC Directive, we shall not be considering this "inter-system" aspect to any great extent. The subject has a long history and there are many textbooks dealing with it.

1.1.1.2 Equipment in isolation

The second approach assumes that the system will operate in an environment which is electromagnetically benign within certain limits, and that its proximity to other sensitive equipment will also be controlled within limits. So for example, most of the time a personal computer will not be operated in the vicinity of a high power radar transmitter, nor will it be put right next to a mobile radio receiving antenna. This allows a very broad set of limits to be placed on both the permissible emissions from a device and on the levels of disturbance within which the device should reasonably be expected to continue operating. These limits are directly related to the class of environment –

domestic, commercial, industrial etc. – for which the device is marketed. The limits and the methods of demonstrating that they have been met form the basis for a set of standards, some aimed at emissions and some at immunity, for the EMC performance of any given product in isolation.

Note that compliance with such standards will not guarantee electromagnetic compatibility under all conditions. Rather, it establishes a probability (hopefully very high) that equipment will not cause interference nor be susceptible to it when operated under *typical* conditions. There will inevitably be some special circumstances under which proper EMC will not be attained – such as operating a computer within the near field of a powerful transmitter – and extra protection measures must be accepted.

1.1.2 The scope of EMC

The principal issues which are addressed by EMC are discussed below. The use of microprocessors in particular has stimulated the upsurge of interest in EMC. These devices are widely responsible for generating radio frequency interference and are themselves susceptible to many interfering phenomena. At the same time, the widespread replacement of metal chassis and cabinets by moulded plastic enclosures has drastically reduced the degree of protection offered to circuits by their housings.

1.1.2.1 Malfunction of systems

Solid state and especially processor-based control systems are taking over many functions which were earlier the preserve of electromechanical or analogue equipment such as relay logic or proportional controllers. Rather than being hard-wired to perform a particular task, programmable electronic systems rely on a digital bus-linked architecture in which many signals are multiplexed onto a single hardware bus under software control. Not only is such a structure more susceptible to interference, because of the low level of energy needed to induce a change of state, but the effects of the interference are impossible to predict; a random pulse may or may not corrupt the operation depending on its timing with respect to the internal clock, the data that is being transferred and the program's execution state. Continuous interference may have no effect as long as it remains below the logic threshold, but when it increases further the processor operation will be completely disrupted. With increasing functional complexity comes the likelihood of system failure in complex and unexpected failure modes.

Clearly the consequences of interference to control systems will depend on the value of the process that is being controlled. In some cases disruption of control may be no more than a nuisance, in others it may be economically damaging or even life threatening. The level of effort that is put into assuring compatibility will depend on the expected consequences of failure.

Phenomena

Electromagnetic phenomena which can be expected to interfere with control systems are

- supply voltage interruptions, dips, surges and fluctuations;
- transient overvoltages on supply, signal and control lines;
- radio frequency fields, both pulsed (radar) and continuous, coupled directly into the equipment or onto its connected cables;
- electrostatic discharge (ESD) from a charged object or person;

- low frequency magnetic or electric fields.

Note that we are not directly concerned with the phenomenon of component damage due to ESD, which is mainly a problem of electronic production. Once the components are assembled into a unit they are protected from such damage unless the design is particularly lax. But an ESD transient can corrupt the operation of a microprocessor or clocked circuit just as a transient coupled into the supply or signal ports can, without actually damaging any components (although this may also occur), and this is properly an EMC phenomenon.

Software

Malfunctions due to faulty software may often be confused with those due to EMI. Especially with real time systems, transient coincidences of external conditions with critical software execution states can cause operational failure which is difficult or impossible to replicate, and may survive development testing to remain latent for years in fielded equipment. The symptoms – system crashes, incorrect operation or faulty data – can be identical to those induced by EMI. In fact you may only be able to distinguish faulty software from poor EMC by characterizing the environment in which the system is installed.

1.1.2.2 Interference with radio reception

Bona fide users of the radio spectrum have a right to expect their use not to be affected by the operation of equipment which is nothing to do with them. Typically, received signal strengths of wanted signals vary from less than a microvolt to more than a millivolt, at the receiver input. If an interfering signal is present on the same channel as the wanted signal then the wanted signal will be obliterated if the interference is of a similar or greater amplitude. The acceptable level of co-channel interference (the "protection factor") is determined by the wanted programme content and by the nature of the interference. Continuous interference on a high fidelity broadcast signal would be unacceptable at very low levels, whereas a communications channel carrying compressed voice signals can tolerate relatively high levels of impulsive or transient interference.

Field strength level

Radiated interference, whether intentional or not, decreases in strength with distance from the source. For radiated fields in free space, the decrease is inversely proportional to the distance provided that the measurement is made in the far field (see section 4.1.3.2 for a discussion of near and far fields). As ground irregularity and clutter increase, the fields will be further reduced because of shadowing, absorption, scattering, divergence and defocussing of the diffracted waves. Annex D of EN 55 011 [86] suggests that for distances greater than 30m over the frequency range 30 to 300MHz, the median field strength varies as $1/d^n$ where n varies from 1.3 for open country to 2.8 for heavily built-up urban areas. An average value of $n = 2.2$ can be taken for approximate estimations; thus increasing the separation by ten times would give a drop in interfering signal strength of 44dB.

Limits for unintentional emissions are based on the acceptable interfering field strength that is present at the receiver – that is, the minimum wanted signal strength for a particular service modified by the protection ratio – when a nominal distance separates it from the emitter. This will not protect the reception of very weak wanted signals nor will it protect against the close proximity of an interfering source, but it will cover the majority of interference cases and this approach is taken in all those standards for emission limits that have been published for commercial equipment by CISPR (see chapter 2). CISPR publication 23 [103] gives an account of how such limits are derived, including the statistical basis for the probability of interference occurring.

Below 30MHz the dominant method of coupling out of the interfering equipment is via its connected cables, and therefore the radiated field limits are translated into equivalent voltage or current levels that, when present on the cables, correspond to a similar level of threat to HF and MF reception.

Malfunction versus spectrum protection

It should be clear from the foregoing discussion that RF emission limits are not determined by the need to guard against malfunction of equipment which is not itself a radio receiver. As discussed in the last section, malfunction requires fairly high energy levels – RF field strengths in the region of 1 – 10 volts per meter for example. Protection of the spectrum for radio use is needed at much lower levels, of the order of 10 – 100 microvolts per metre – ten to a hundred thousand times lower. RF incompatibility between two pieces of equipment neither of which intentionally uses the radio spectrum is very rare. Normally, equipment immunity is required from the local fields of intentional radio transmitters, and unintentional emissions must be limited to protect the operation of intentional radio receivers. The two principal EMC aspects of emissions and immunity therefore address two different issues.

Free radiation frequencies

Certain types of equipment generate high levels of RF energy but use it for purposes other than communication. Medical diathermy and RF heating apparatus are examples. To place blanket emissions limits on this equipment would be unrealistic. In fact, the International Telecommunications Union (ITU) has designated a number of frequencies specifically for this purpose, and equipment using only these frequencies (colloquially known as the "free radiation" frequencies) is not subject to emission restrictions. Table 1.1 lists these frequencies.

Centre frequency, MHz	Frequency range, MHz	
6.780	6.765 - 6.795	*
13.560	13.553 - 13.567	
27.120	26.957 - 27.283	
40.680	40.66 - 40.70	
433.920	433.05 - 434.79	*
2,450	2,400 - 2,500	
5,800	5,725 - 5,875	
24,125	24,000 - 24,250	
61,250	61,000 - 61,500	*
122,500	122,000 - 123,000	*
245,000	244,000 - 246,000	*

* : maximum radiation limit under consideration, use subject to special authorization
Source: EN55011

Table 1.1 ITU designated industrial, scientific and medical free radiation frequencies

Co-channel interference

A further problem with radio communications, often regarded as an EMC issue although it will not be treated in this book, is the problem of co-channel interference from unwanted transmissions. This is caused when two radio systems are authorised to use the same frequency on the basis that there is sufficient distance between the systems, but abnormal propagation conditions increase the signal strengths to the point at which interference is noticeable. This is essentially an issue of spectrum utilisation.

A transmitted signal may also overload the input stages of a nearby receiver which is tuned to a different frequency and cause desensitisation or distortion of the wanted signal. Transmitter outputs themselves will have spurious frequency components present as well as the authorised frequency, and transmitter type approval has to set limits on these spurious levels.

1.1.2.3 Disturbances on the mains supply

Mains electricity suffers a variety of disturbing effects during its distribution. These may be caused by sources in the supply network or by other users, or by other loads within the same installation. A pure, uninterrupted supply would not be cost effective; the balance between the cost of the supply and its quality is determined by national regulatory requirements, tempered by the experience of the supply utilities. Typical disturbances are:

- *voltage variations*: the distribution network has a finite source impedance and varying loads will affect the terminal voltage. Including voltage drops within the customer's premises, an allowance of ±10% on the nominal voltage will cover normal variations in the UK; proposed limits for all CENELEC countries (see pr EN 50 093, section 2.3.3) are +12%, –15%.
- *voltage fluctuations*: short-term (sub-second) fluctuations with quite small amplitudes are annoyingly perceptible on electric lighting, though they are comfortably ignored by electronic power supply circuits. Generation of flicker by high power load switching is subject to regulatory control.
- *voltage interruptions*: faults on power distribution systems cause almost 100% voltage drops but are cleared quickly and automatically by protection devices, and throughout the rest of the distribution system the voltage immediately recovers. Most consumers therefore see a short voltage dip. The frequency of occurrence of such dips depends on location and seasonal factors.
- *waveform distortion*: at source, the AC mains is generated as a pure sine wave but the reactive impedance of the distribution network together with the harmonic currents drawn by non-linear loads causes voltage distortion. Power converters and electronic power supplies are important contributors to non-linear loading. Harmonic distortion may actually be worse at points remote from the non-linear load because of resonances in the network components. Not only must non-linear harmonic currents be limited but equipment should be capable of operating with up to 10% total harmonic distortion in the supply waveform.
- *transients and surges*: switching operations generate transients of a few hundred volts as a result of current interruption in an inductive circuit. These transients normally occur in bursts and have risetimes of no more than a few nanoseconds, although the finite bandwidth of the distribution network will

quickly attenuate all but local sources. Rarer high amplitude spikes in excess of 2kV may be observed due to fault conditions. Even higher voltage surges due to lightning strikes occur, most frequently on exposed overhead line distribution systems in rural areas.

All these sources of disturbance can cause malfunction in systems and equipment that do not have adequate immunity.

Mains signalling

A further source of incompatibility arises from the use of the mains distribution network as a telecommunications medium, or mains signalling (MS). MS superimposes signals on the mains in the frequency band from 3kHz to 150kHz and is used both by the supply industry itself and by consumers. Unfortunately this is also the frequency band in which electronic power converters – not just switch-mode power supplies, but variable speed motor drives, induction heaters, fluorescent lamp inverters and similar products – operate to their best efficiency. There are at present no pan-European standards which regulate conducted emissions on the mains below 150kHz, although BS 6839 part 1 (EN 50 065 - 1) [109] sets the frequency allocations and output and interference limits for MS equipment itself. The German RF emission standard VDE 0871 extends down to 9kHz for some classes of equipment. Overall, compatibility problems between MS systems and such power conversion equipment can be expected to increase.

1.1.2.4 Other EMC issues

The issues discussed above are those which directly affect product design to meet commercial EMC requirements, but there are two other aspects which should be mentioned briefly.

EEDs and flammable atmospheres

The first is the hazard of ignition of flammable atmospheres in petrochemical plant, or the detonation of electro-explosive devices in places such as quarries, due to incident RF energy. A strong electromagnetic field will induce currents in large metal structures which behave as receiving antennas. A spark will occur if two such structures are in intermittent contact or are separated. If flammable vapour is present at the location of the spark, and if the spark has sufficient energy, the vapour will be ignited. Different vapours have different minimum ignition energies, hydrogen/air being the most sensitive. The energy present in the spark depends on the field strength, and hence on the distance from the transmitter, and on the antenna efficiency of the metal structure. BS 6656 [107] discusses the nature of the hazard and presents guidelines for its mitigation.

Similarly, electro-explosive devices (EEDs) are typically connected to their source of power for detonation by a long wire, which can behave as an antenna. Currents induced in it by a nearby transmitter could cause the charges to explode prematurely if the field was strong enough. As with ignition of flammable atmospheres, the risk of premature detonation depends on the separation distance from the transmitter and the efficiency of the receiving wire. EEDs can if necessary be filtered to reduce their susceptibility to RF energy. BS 6657 [108] discusses the hazard to EEDs.

Data security

The second aspect of EMC is the security of confidential data. Low level RF emissions from data-processing equipment may be modulated with the information that the equipment is carrying – for instance, the video signal that is fed to the screen of a VDU. These signals could be detected by third parties with sensitive equipment located outside a secure area and demodulated for their own purposes, thus compromising the security of the overall system. This threat is already well recognised by government agencies and specifications for emission control, under the Tempest scheme, have been established for many years. Commercial institutions, particularly in the finance sector, are now beginning to become aware of the problem.

1.1.3 The compatibility gap

The increasing susceptibility of electronic equipment to electromagnetic influences is being paralleled by an increasing pollution of the electromagnetic environment. Susceptibility is a function partly of the adoption of VLSI technology in the form of microprocessors, both to achieve new tasks and for tasks that were previously tackled by electromechanical or analogue means, and the accompanying reduction in the energy required of potentially disturbing factors. It is also a function of the increased penetration of radio communications, and the greater opportunities for interference to radio reception that result from the co-location of unintentional emitters and radio receivers.

At the same time more radio communications mean more transmitters and an increase in the average RF field strengths to which equipment is exposed. Also, the proliferation of digital electronics means an increase in low-level emissions which affect radio reception, a phenomenon which has been aptly described as a form of electromagnetic "smog".

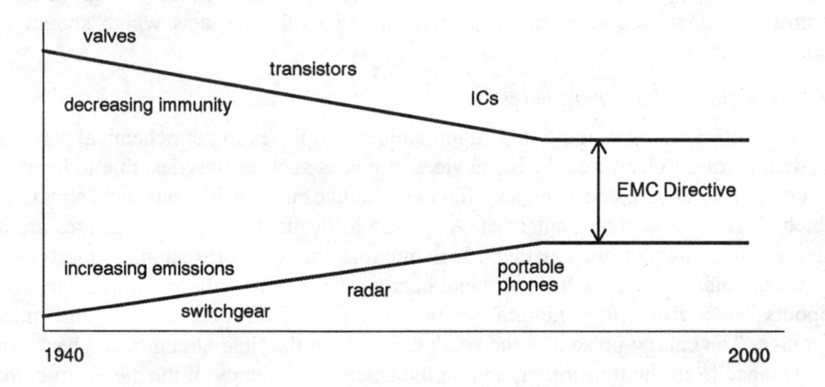

Figure 1.1 The EMC gap

These concepts can be graphically presented in the form of a narrowing electromagnetic compatibility gap, as in Figure 1.1. This "gap" is of course conceptual rather than absolute, since phenomena defined as emissions and those defined as immunity do not mutually interact except in rare cases. But the maintenance of some artificially-defined gap between immunity and emissions is the purpose of the application of EMC standards, and is the entirely worthy aim of the enforcement of the EMC Directive.

1.2 The EMC Directive

The relaxed EMC regime that had hitherto existed throughout most of Europe has now been totally overturned with the adoption on 1st January 1992 by the European Commission of the EMC Directive, 89/336/EEC [112]. This is widely regarded to be "the most comprehensive, complex and possibly contentious Directive ever to emanate from Brussels" [22]. The remainder of this chapter examines the provisions of the Directive and how manufacturers will need to go about complying with it.

1.2.1 The new approach directives

The Single European Market is intended to be in place and operational by the end of 1992. Of the various aims of the creation of the Single Market, the free movement of goods between European states† is fundamental. All member states impose standards and obligations on the manufacture of goods in the interests of quality, safety, consumer protection and so forth. Because of detailed differences in procedures and requirements, these act as technical barriers to trade, fragmenting the European market and increasing costs because manufacturers have to modify their products for different national markets.

For many years the EC tried to remove these barriers by proposing Directives which gave the detailed requirements that products had to satisfy before they could be freely marketed throughout the Community, but this proved difficult because of the detailed nature of each Directive and the need for unanimity before it could be adopted. In 1985 the Council of Ministers adopted a resolution setting out a "New Approach to Technical Harmonisation and Standards".

Under the "new approach", directives are limited to setting out the essential requirements which must be satisfied before products may be marketed anywhere within the EC. The technical detail is provided by standards drawn up by the European standards bodies CEN, CENELEC and ETSI. Compliance with these standards will demonstrate compliance with the essential requirements of each Directive. All products covered by each Directive must meet its essential requirements, but all products which do comply, and are labelled as such, may be circulated freely within the Community; no member state can refuse them entry on technical grounds. Decisions on new approach Directives are taken by qualified majority voting, eliminating the need for unanimity and speeding up the process of adoption.

Contents

A new approach Directive contains the following elements [115]:

- the *scope* of the Directive
- a statement of the *essential requirements*
- the *methods of satisfying* the essential requirements
- how *evidence of conformity* will be provided
- what *transitional arrangements* may be allowed
- a statement confirming entitlement to *free circulation*
- a *safeguard procedure*, to allow Member States to require a product to be withdrawn from the market if it does not satisfy the essential requirements

It is the responsibility of the European Commission to put forward to the Council of

† Appendix E lists the EC Member States.

Ministers proposals for new Directives. Directorate-General III of the Commission has the overall responsibility for the EMC Directive. The actual decision on whether or not to adopt a proposed Directive is taken by the Council of Ministers, by a qualified majority of 54 out of 76 votes (the UK, France, Germany and Italy each have ten votes; Spain has eight votes; Belgium, Greece, The Netherlands and Portugal each have five votes; Denmark and Ireland have three votes and Luxembourg has two votes). Texts of Directives proposed or adopted are published in the *Official Journal of the European Communities*. Consultation on draft Directives is typically carried out through European representative bodies and in working parties of governmental experts.

Other Directives

Apart from the EMC Directive, other new approach Directives adopted or under discussion at the time of writing which may affect some sectors of the electrical and electronic engineering industry are

- Toy safety
- Non-automatic weighing machines
- Active medical devices
- Active implantable electromedical devices
- Machinery safety

1.2.2 Background to the legislation

In the UK, previous legislation on EMC has been limited in scope to radio communications. Section 10 of the Wireless Telegraphy Act 1949 enables regulations to be made for the purpose of controlling both radio and non-radio equipment which might interfere with radio communications. These regulations have taken the form of various Statutory Instruments (SIs) which cover interference emissions from spark ignition systems, electromedical apparatus, RF heating, household appliances, fluorescent lights and CB radio. The SIs invoke British Standards which are closely aligned with international and European standards (see chapter 2).

The power exists to make regulations regarding the immunity to interference of radio equipment but this has so far not been used.

At the European level various Directives have been adopted over the years, again to control emissions from specific types of equipment. Directive 72-245 EEC, adopted in June 1972, regulates interference produced by spark ignition engines in motor vehicles. Directives 76-889 EEC and 76-890 EEC, amended by various other subsequent Directives, apply to interference from household appliances and portable tools, and fluorescent lamps and luminaires. These latter two will be superseded and repealed by the new EMC Directive. Each member state is required to implement the provisions of these Directives in its national legislation, as described above for the UK.

This previous legislation is not comparable in scope to the EMC Directive, which covers far more than just interference to radio equipment, and extends to include immunity as well as emissions.

1.2.3 Scope, requirements and exceptions

The EMC Directive, 89/336/EEC, applies to apparatus which is liable to cause electromagnetic disturbance or which is itself liable to be affected by such disturbance. "Apparatus" is defined as all electrical and electronic appliances, equipment and installations. Essentially, anything which is powered by electricity is covered, regardless of whether the power source is the public supply mains, a battery source or a specialised supply.

An electromagnetic disturbance is any electromagnetic phenomenon which may degrade performance, without regard to frequency or method of coupling. Thus radiated emissions as well as those conducted along cables; and immunity from EM fields, mains disturbances, conducted transients and RF, electrostatic discharge and lightning surges are all covered. *No* specific phenomena are *excluded* from the Directive's scope.

Essential requirements

The essential requirements of the Directive (Article 4) are that the apparatus shall be so constructed that:

- the electromagnetic disturbance it generates does not exceed a level allowing radio and telecommunications equipment and other apparatus to operate as intended;
- the apparatus has an adequate level of intrinsic immunity to electromagnetic disturbance to enable it to operate as intended.

The intention is to protect the operation not only of radio and telecommunications equipment but any equipment which might be susceptible to EM disturbances, such as information technology or control equipment. At the same time, all equipment must be able to function correctly in whatever environment it might reasonably be expected to occupy. Notwithstanding these requirements, any member state has the right to apply special measures with regard to the taking into service of apparatus, to overcome existing or predicted EMC problems at a specific site or to protect the public telecommunications and safety services. Compliance with the essential requirements will be demonstrated by two main paths, that is certification to harmonized standards or via a technical construction file. These are discussed in section 1.3.

1.2.3.1 Sale and use of products

The Directive applies to all apparatus that is placed on the market or taken into service. The definitions of these two conditions do not appear within the text of the Directive but have been the subject of an interpretative document issued by the Commission [117].

Placed on the market

The "market" means the market in any or all of the EC member states; products which are found to comply within one state are automatically deemed to comply within all others. "Placing on the market" means the *first* making available of the product within the EC, so that the Directive covers only new products manufactured within the EC, but both new and used products imported from a third country. Products sold second hand within the EC are outside its scope. Where a product passes through a chain of distribution before reaching the final user, it is the passing of the product from the manufacturer into the distribution chain which constitutes placing on the market. If the product is manufactured in or imported into the EC for subsequent export to a third country, it has not been placed on the market.

The Directive applies to each individual item of a product type regardless of when it was designed, and whether it is a one-off or high volume product. Thus items from a product line that was launched at any time before 1992 must comply with the provisions of the Directive after 1st January 1992 – but note the effect of the transitional period, discussed in section 1.3.3. Put another way, there is no "grandfather clause" which exempts designs that were current before the Directive took effect. However, products already *in use* before 1st January 1992 do not have to comply retrospectively.

Taken into service

"Taking into service" means the first *use* of a product in the EC by its final user. If the product is used without being placed on the market, if for example the manufacturer is also the end user, then the protection requirements of the Directive still apply. This means that sanctions are still available in each member state to prevent the product from being used if it does not comply with the essential requirements or if it causes an actual or potential interference problem. On the other hand, it should not need to go through the conformity assessment procedures to demonstrate compliance (article 10, which describes these procedures, makes no mention of taking into service). Thus an item of special test gear built up by a lab technician for use within the company's design department must still be designed and tested to conform to EMC standards, but should not need to follow the procedure for applying the CE mark.

If the manufacturer resides outside the EC, then the responsibility for certifying compliance with the Directive rests with the person placing the product on the market for the first time within the EC, i.e. the manufacturer's authorized representative or the importer. Any person who produces a new finished product from already existing finished products, such as a system builder, is considered to be the manufacturer of the new finished product.

1.2.3.2 Exceptions

There are a few specific exceptions from the scope of the Directive, but these are not such as to offer cause for much relief. Self-built amateur radio apparatus (but not CB equipment) is specifically excluded. The only other exclusions are for those types of apparatus which are subject to EMC requirements in other Directives. At the time of writing these are motor vehicle ignition systems (emissions) and non-automatic weighing machines (immunity); the complementary EMC aspects of these products *are* covered by the EMC Directive. Medical devices are fully covered by other Directives.

Military equipment is excluded from the scope of the EMC Directive as a result of an exclusion clause in the Treaty of Rome, but equipment which has a dual military/civil use will be covered when it is placed on the civilian market. Educational electronic equipment intended for the study of electromagnetic phenomena need not meet the essential requirements – since its whole purpose is deliberately to emit or be susceptible to interference – provided that its user ensures that it does not affect equipment installed elsewhere.

Components

The question of when does a "component" (which is not within the scope of the Directive) become "apparatus" (which is) remains problematical. The Commission's interpretative document defines a component to be "any item which is used in the composition of an apparatus and which is not itself an apparatus with an intrinsic function intended for the final consumer". Thus individual small parts such as ICs and resistors are definitely outside the Directive. A component may be more complex provided that it does not have an intrinsic function and its only purpose is to be incorporated inside an apparatus, but the manufacturer of such a component must indicate to the equipment manufacturer how to use and incorporate it. The distinction is important for manufacturers of board-level products and other sub-assemblies which may appear to have an intrinsic function and are marketed separately, yet cannot be used separately from the apparatus in which they will be installed. The Commission has indicated that it regards sub-assemblies which are to be tested as part of a larger apparatus as outside the Directive's scope.

At the other extreme, the Directive specifically does *not* apply to apparatus which is not liable to cause or be susceptible to interference. No guidance is given as to how to assess such lack of liability, but for instance a battery operated torch or a domestic electric fire would clearly fall under this heading – although the same could not be said for example of a battery operated device containing a motor.

1.2.4 The CE mark and the declaration of conformity

The manufacturer or his authorized representative is required to attest that the protection requirements of the Directive have been met. This requires two things:

- he issues a declaration of conformity which must be kept available to the enforcement authority for ten years following the placing of the apparatus on the market;
- he affixes the CE mark to the apparatus, or to its packaging, instructions or guarantee certificate.

A draft Council Regulation concerning the affixing and use of the CE mark has been published for comment [119]. The mark consists of the letters CE as shown in Figure 1.2 together with the year in which the mark was affixed, and also where necessary the identifying number of the notified body which issued the EC type examination certificate (this latter is at present applicable only to telecommunications terminal equipment and radio transmitters). The mark should be at least 5mm in height and be affixed "visibly, legibly and indelibly" but its method of fixture is not otherwise specified. Affixing this mark indicates conformity not only with the EMC Directive but also with the requirements of any other new approach Directives relevant to the product – for instance, an electrical toy with the CE mark indicates compliance both with the Toy Safety Directive and the EMC Directive. This provision, together with the requirement for a date mark, is likely to cause many difficulties for manufacturers who have to comply with several Directives which have different implementation dates and transitional arrangements, and in practice the requirement for a date mark may be negotiated away.

Figure 1.2 The CE mark

The EC declaration of conformity is required whether the manufacturer self-certifies to harmonized standards or follows the technical file route (section 1.3). It must include the following components:

- a description of the apparatus to which it refers;

- a reference to the specifications under which conformity is declared, and where appropriate to the national measures implemented to ensure conformity;
- an identification of the signatory empowered to bind the manufacturer or his authorized representative;
- where appropriate, reference to the EC type examination certificate (see above).

1.2.4.1 Description

The description of the apparatus should be straightforward; assuming the equipment has a type number, then reference to this type number (provided that supporting documentation is available) should be sufficient. Difficulties arise when the type is subjected to revision or modification. At what stage do modifications or updates result in a new piece of equipment that would require re-certification? If the declaration of conformity refers to the Widget 2000 with software version 1.0 launched in 1993, does it continue to refer to the Widget 2000S of 1996 with version 3.2? The sensible approach would be to determine whether the modifications had affected the EMC performance and if so, re-issue the declaration for the new product; but this will require that you re-test the modifications, with the attendant cost penalties, or you exercise some engineering judgement as to whether a minor change will affect performance. No general guidance can be given on this point, but it should be clear that the breadth of the EMC requirements means that very few modifications will have absolutely no effect on a product's EMC performance.

1.2.4.2 Signatory

The empowered signatory will not necessarily be competent to judge the technicalities of what is being declared. Normally this will be one of the directors of the manufacturing or importing company. In small companies the technical director will probably be close enough to the product in question to understand the detail of its EMC performance, but in medium or large-scale enterprises the directors will increasingly have to rely on the technical advice of their product development and manufacturing engineers and/or the EMC test and management personnel. Such companies will have to define clearly the levels of responsibility that exist for each person involved in making the declaration.

1.2.4.3 Specifications

The reference to the specifications under which conformity is declared does not necessarily mean that you have to *test* to these specifications. Three possibilities are apparent. Firstly, you may deem that the product intrinsically meets the requirements of the Directive and does not need testing. An example might be a simple linear unregulated stand-alone power supply which is below the power level at which harmonic currents are controlled. Provided that you can convince the signatory that this is a competent engineering judgement then there is no reason not to make a valid declaration on this basis. Most electronic products will not be able to follow this option.

Secondly, you may be able to make a declaration based on existing test results. If for example you have already been conforming to existing standards such as FCC or VDE for emissions and an internal immunity standard based on IEC 801, then you may be confident enough to state that the product will meet the appropriate harmonized standards without further testing, or with only partial testing.

The final option is to test fully to harmonized standards or to choose the technical file route. For a sophisticated product either of these will be lengthy and expensive and may involve some complex judgements as to what tests to apply, especially if appropriate standards are not available. For new products though, testing will be essential. A fourth option, of course, is not to test at all; just make the declaration, stick on the CE mark and hope that nobody ever notices. A reputable company, of course, won't take this route, but the possibility of competitors doing so may be a factor in assessing your market position.

1.2.5 Manufacturing quality assessment

The Directive covers every individual, physically existing finished product, but it would be impractical to test every item in series production fully for all the EMC characteristics that it must exhibit. The Directive itself is silent on quality assurance procedures, although the Commission's interpretative document reminds manufacturers of their responsibility to ensure ongoing conformity. The conformity assessment procedures for all the technical harmonization Directives are contained in Council Decision 90/683/EEC [118]. This document contains a range of modules which may be applied in the case of each specific Directive. The EMC Directive will use module A of this Decision.

This module (referred to as internal production control) requires the manufacturer (or his authorized representative) to establish and keep for ten years *after the last product has been manufactured*, at the disposal of the national authorities, a set of technical documentation†. The purpose of this documentation is to enable the conformity of the product with the Directive's requirements to be assessed, and it will cover for such purpose the design, manufacture and operation of the product. It will contain so far as is relevant

- a general description of the product,
- design and manufacturing drawings,
- descriptions and explanations necessary for the understanding of such drawings and the product's operation,
- a list of the standards applied, and descriptions of solutions adopted to meet the essential requirements when standards have not been applied,
- results of design calculations and examinations carried out,
- test reports.

A copy of the declaration of conformity will be kept with this documentation.

Note that this documentation is generally equivalent to the design documentation that all manufacturers should keep for each of their products, especially if they are operating under a design quality system such as ISO9000, except that it refers only (as far as the EMC Directive is concerned) to the EMC aspects of the product. It is not held publicly, only "at the disposal of" the national authorities, and therefore considerations of confidentiality are not relevant.

1.2.5.1 Production control

Having established the technical documentation, Module A goes on to require that "the manufacturer shall take all measures necessary in order that the manufacturing process shall ensure compliance of the manufactured products with the technical

† Note that this is not the same as the technical construction file, for which see section 1.3.2.

documentation ... and with the requirements of the directive that apply to them". No specific means of determining what these measures might be are mentioned in either the Directive or the Council Decision.

CISPR sampling schemes

For many years before the adoption of the EMC Directive, the standards committee CISPR (see section 2.1.1) had recognised the need for some form of production quality testing, and had incorporated sampling schemes into the emission standards which form the basis of EN55011, EN55014 and EN55022. The purpose of these schemes is to ensure that at least 80% of series production complies with the limits with an 80% confidence level, the so-called 80/80 rule. Practically, to comply fully with the 80/80 rule the manufacturer has to aim at about 95% of the products being in compliance with the specified limit.

The first scheme requires measurements of the actual emission levels from between 3 and 12 identical items, from which the mean and standard deviation are derived. The limit levels are then expressed in the form

$$L \geq \overline{X} + k \cdot S_n \qquad (1.1)$$

where $\overline{X}$ is the arithmetic mean and S_n the standard deviation of the measured emission levels, and k is a constant derived from the non-central t-distribution between 2.04 and 1.2 depending on sample size

If the emission levels are similar between items (a low value of S_n) then a small margin below the limit is needed; if they are highly variable, then a large margin is needed. This sampling method can only be applied to emissions measurements and cannot be used for immunity.

A second scheme which is applicable to both emissions and immunity is based on recording test failures over a sample of units. Compliance is judged from the condition that the number of units with an immunity level below the specified limit, or that exceed the emissions limits, may not exceed c in a sample of size n: this test is based on the

n	7	14	20	26	32
c	0	1	2	3	4

binomial distribution and produces the same result as the first, in accordance with the 80/80 rule.

As well as the above sampling schemes, published EN standards also allow a single test to be made on one item only, but then advise that subsequent tests are necessary from time to time on samples taken at random from production. The banning of sales is to occur only after tests have been carried out in accordance with one or other sampling scheme.

1.2.6 Systems and installations

A particularly contentious area is how the Directive applies to two or more separate pieces of apparatus sold together or installed and operating together. It is clear that the Directive applies in principle to systems and installations. The Commission's interpretative document [117] defines a system as several items of apparatus combined to fulfil a specific objective and intended to be placed on the market as a single functional unit. An installation is several combined items of apparatus or systems put together at a given place to fulfil a specific objective but not intended to be placed on the market as a single functional unit. Therefore a typical system would be a personal computer workstation comprising the PC, monitor, keyboard, printer and any other peripherals. If the units were to be sold separately they would have to be tested and certified separately; if they were to be sold as a single package then they would have to be tested and certified as a package.

Any other combination of items of apparatus, not initially intended to be placed on the market together, is considered to be not a system but an installation. Examples of this would appear to be computer suites, telephone exchanges, electricity substations or television studios. Each item of apparatus in the installation is subject to the provisions of the Directive individually, under the specified installation conditions.

As far as it goes, this interpretation is useful, in that it allows testing and certification of installations to proceed on the basis that each component of the installation will meet the requirements on its own. The difficulty of testing large installations in situ against standards that were never designed for them is largely avoided. Also, if an installation uses large numbers of similar or identical components then only one of these needs to be actually tested.

Large systems

The definition unfortunately does not help system builders who will be "placing on the market" – i.e. supplying to their customer on contract – a single installation, made up of separate items of apparatus but actually sold as one functional unit. Many industrial, commercial and public utility contracts fall into this category. According to the published interpretation, the overall installation would be regarded as a system and therefore should comply as a package. As it stands at present, there are no standards which specifically cover large systems, i.e. ones for which testing on a test site is impractical, although some emissions standards do allow measurements in situ. These measurements are themselves questionable because of the difficulty of distinguishing external interference at the measurement position from that due to the installation, and because the variability of the physical installation conditions introduces reflections and standing waves which distort the measurement. There are no provisions for large systems in the immunity standards. Therefore the only compliance route available to system builders is the technical construction file (section 1.3.2), but guidance as to how to interpret the Directive's essential requirements in these cases is lacking. The principle dilemma of applying the Directive to complete installations is that to make legally relevant tests is difficult, but the nature of EMC phenomena is such that to test only the constituent parts without reference to their interconnection is meaningless.

A draft standard pr ETS 300127 [96], written around large telecommunication systems, has been proposed to remedy the deficiencies in radiated emissions testing. It allows testing of a minimum representative system on an open area test site; such a system contains at least one of each sub-unit type which will be included in operational systems, and comprises the minimum configuration (including interface lines) of any system that is offered for sale. The system cable configuration is specifically addressed in this standard. The test on the representative system is used for demonstration of compliance. The EMC performance of new functional modules can be assessed against those that they replace. The representative system continues to conform when a functional module has been replaced by a new functional module which exhibits emissions similar to or less than the original unit. This approach has much to recommend it outside the telecomms sector, and system builders in other areas would be well advised to study this standard (possibly also to be published as EN 50 098). A similarly worded section has been proposed as a revision to EN 55 022.

It is possible that one-off installations which are manufactured to a customer's individual requirements, rather than being available off the shelf, could be treated under the provisions of taking into service. In this case it might not be subject to the need for testing and certification, but would still have to meet the protection requirements, and the user would be responsible for any special measures if the installation caused excessive interference.

1.2.7 Enforcement and sanctions

Member states cannot impede for EMC reasons the free circulation of apparatus covered by the Directive which meets its requirements when properly installed and maintained, and used for its intended purpose. They must presume that apparatus which bears the CE mark and conforms to the relevant standards, or for which a technical construction file exists, does in fact comply with the protection requirements unless there is evidence to the contrary.

On the other hand, member states are required to ensure that equipment which is found not to comply is not placed on the market or taken into service, and to take appropriate measures to withdraw non-compliant apparatus from the market. Legislation which translates the Directive's requirements into national law in each member state was required to be in place by 1st July 1991, although in the UK at least this timetable has slipped because of the need for clarification on some points by the Commission, and the implementing legislation is unlikely to be in place before mid-92. In English law it is customary to interpret comprehensively the requirements of EC Directives, whereas in other European countries the regulations tend to be more general. Even so, with the exception of Denmark which transposed the EMC Directive into law in April 1991, the other European countries are also delaying transposition until the situation is clearer.

In the UK the proposed enforcement regime [113][52] will include the issue of "prohibition notices" which will prohibit the supply or use of specified equipment which the enforcement authorities believe does not comply with the EMC requirements. The notice may or may not have immediate effect, depending on the urgency of the situation; an appeal procedure will allow persons on whom a prohibition notice is served to make representations for it to be revoked. Enforcement authorities will also be able to apply to a court for forfeiture of apparatus, with its consequent destruction, modification or disposal, and officers of the enforcement authorities may be empowered to enter premises and inspect or seize apparatus and documents.

1.2.7.1 Offences

The EMC legislation will include criminal sanctions. But because of the difficulty of judging whether or not apparatus does comply with the requirements, the UK legislators do not propose to create an absolute criminal offence of supplying or using non-compliant equipment. Users and retailers cannot normally be expected to know whether or not the apparatus in question is non-compliant. Criminal offences on other fronts will be necessary, for instance to guard against misuse of the CE mark or the provision of false or misleading information to a competent or notified body, and to penalise breaches of prohibition notices.

1.2.7.2 Practice

Two important questions are: how will enforcement be operated in practice, and will the Directive be enforced equally in all Member States (the so-called "level playing field"). These questions are directly related to the resources that national governments are prepared to devote to the task. The UK DTI has indicated that its enforcement efforts will be complaint-driven, although it is not entirely clear whether this means that it will only investigate interference complaints arising from actual use of apparatus, or whether it will be open to complaints that apparatus does not conform to the Directive's requirements regardless of whether there is a problem in its use. If the latter, then it is likely that a common source of complaint will be from companies testing samples of their competitors' equipment and, if they find that it does not comply, "shopping" them to the authorities.

On the other hand, the German authorities have stated that it will be necessary to gain information from the market in the form of random spot checks in order to react to violations [71]. Germany already has a strong regime for the control of RF emissions in the form of the mandatory VDE standards, and these are stricter than the EN standards which will be used to demonstrate compliance with the Directive. The Germans will be concerned that the Directive might dilute the effectiveness of their existing regime and will therefore be insisting that it is thoroughly enforced.

It is apparent that differences in enforcement practices within the various Member States will work contrary to the stated intent of the Directive, which is to reduce technical barriers to trade. Article 9 of the Directive requires that "where a Member State ascertains that apparatus accompanied by (a means of attestation) does not comply with the protection requirements ...., it shall take all appropriate measures to withdraw the apparatus from the market, prohibit its placing on the market or restrict its free movement" and shall immediately inform the Commission of any such measure. If the Commission finds, after consultation, that the action is justified, it will inform all other Member States. The competent Member State shall then take appropriate action against the author of the attestation. Therefore any Member State can take immediate action to prohibit an offending apparatus from its own market, but sanctions against the company that put the apparatus on the market in another Member State are dependent on the deliberations of the Commission and on the enforcement practices of the latter Member State.

1.3 Compliance with the Directive

Of themselves, the essential requirements are too generalized to enable manufacturers to declare that their product has met them directly. So Article 10 of the Directive provides alternative routes (Figure 1.3) for manufacturers to achieve compliance with them.

1.3.1 Self certification

The route which is expected to be followed by most manufacturers is self certification to harmonized standards. Harmonized standards are those CENELEC or ETSI standards which have been announced in the *Official Journal of the European Communities* (OJEC). In the UK these are published as dual-numbered BS and EN standards.

The Directive also allows certification against national standards which cover areas

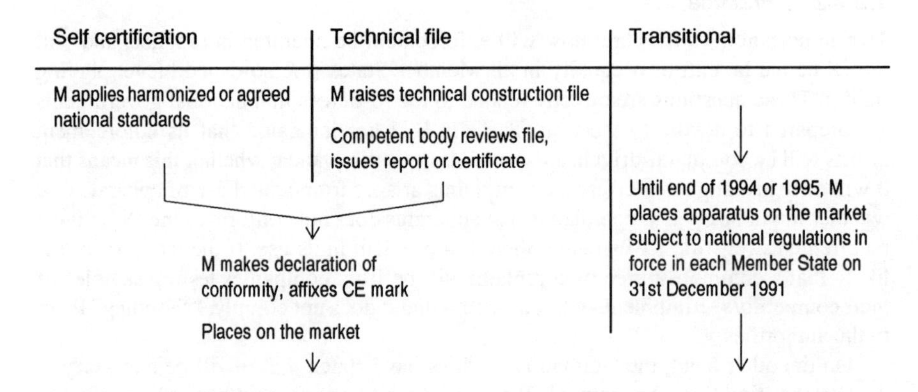

M: manufacturer, authorized representative within the EC, or person who places apparatus on the EC market

Figure 1.3 Routes to compliance

for which no harmonized standards exist, provided that the text of these standards has been notified to the Commission and it has been agreed that they satisfy the Directive's requirements. The reference numbers of these standards must also be published in the OJEC. In practice, the Commission is trying to discourage notification of national standards.

The potential advantage of certifying against standards from the manufacturer's point of view is that there is no mandatory requirement for testing by an independent test house. The only requirement is that the manufacturer makes a declaration of conformity (see section 1.2.4) which references the standards against which compliance is claimed. Of course the manufacturer will normally need to test the product to assure himself that it actually does meet the requirements of the standards, but this could be done in house. Many firms will not have sufficient expertise or facilities in house to do this testing, and will therefore have no choice but to take the product to an independent test house. This is discussed further in section 1.3.5. But the long term aim ought to be to integrate the EMC design and test expertise within the rest of the development or quality department, and to decide which standards apply to the product range, so that the prospect of self certification for EMC is no more daunting than the responsibility of functionally testing a product before shipping it.

1.3.2 The technical construction file

The second route available to achieve compliance is for the manufacturer or importer to generate a Technical Construction File. This is to be held at the disposal of the relevant competent authorities as soon as the apparatus is placed on the market and for ten years thereafter. The Directive specifies that the technical construction file should describe the apparatus, set out the procedures used to ensure conformity with the protection requirements and should contain a technical report or certificate obtained from a competent body.

The purpose of the technical construction file route is to allow compliance with the essential requirements of the Directive to be demonstrated when harmonized or agreed national standards do not exist, or exist only in part, or if the manufacturer chooses not to apply existing standards for his own reasons. Since the generic standards are intended to cover the first two of these cases, the likely useage of this route will be under the following circumstances:

- when existing standards cannot be applied because of the nature of the apparatus or because it incorporates advanced technology which is beyond their breadth of concept;
- when testing would be impractical because of the size or extent of the apparatus, or because of the existence of many fundamentally similar installations;
- when the apparatus is so simple that it is clear that no testing is necessary;
- when the apparatus has already been tested to standards that have not been harmonized or agreed but which are nevertheless believed to meet the essential requirements.

1.3.2.1 Contents

What is required of the technical file's contents are not clear, and the UK DTI is producing a guidance document to clarify the expected level of detail. It would seem reasonable that as well as an identification and general description, any constructional and circuit measures which affect the EMC of the product should be fully described. An explanation and justification for why the technical file route was chosen should be included. This may affect the actual contents.

The technical file may or may not contain test data. The critical item is the technical report or certificate issued by a competent body, and this is what distinguishes this route from the previous one. Essentially, you are required to get an independent qualified opinion on the validity of your belief that the product meets the essential requirements. The competent body should review the technical file to check the rationale for the product's EMC, and the testing that has been done (if any). Either a report or certificate may be issued, both having equivalent weight; and if you have been able to partially apply harmonized standards then this document need only certify conformity with those aspects not covered by these standards.

1.3.2.2 Technical file versus standards

A potentially frequent use of this route would therefore be to test emissions to a harmonized standard but to decide that immunity needed a more product-oriented approach, and to use the technical construction file to certify this aspect. Alternatively, you may decide that an unwarranted amount of effort would be expended in testing against phenomena which experience indicates would not cause problems in real applications. If you can persuade the competent body that this is indeed the case then the technical file and its associated report need merely state this to be so.

Although the technical file route is for use where existing standards are inapplicable, in practice the competent body who issues the report or certificate will have regard to existing standards, methods of measurement and limits in order to judge whether the equipment meets the essential requirements. A close working relationship between the manufacturer and the competent body he chooses will be needed. The expertise and qualifications needed of a competent body are discussed in section 1.3.5.

1.3.2.3 Maintenance of the technical file

The phrase "hold at the disposal of" (the competent authorities) needs further clarification. This may mean the authority in the manufacturer's own Member State, or the authorities in all member states throughout the EC. Is the file made available only on specific request related to a particular apparatus, or are all files to be made available on general request? How does the national authority judge the file if queries arise, and what provisions for discussion and appeal are there? At what point does a new file for apparatus that has undergone modifications need to be started? These points are likely only to be resolved through time and the actual application of the Directive.

1.3.3 The transition period

The EMC Directive as it was adopted in 1989 contained a third, transitional route to compliance. This allowed for apparatus to continue to be governed by "national arrangements in force" on the date of adoption, for a period of one year until 31st December 1992. In practice, this meant that many classes of equipment would continue to be unregulated, especially as regards immunity, in most countries throughout the EC. They could not bear the CE mark and would not therefore benefit from the automatic right of free circulation that this would confer, but neither would they be restricted from being placed on the market as long as they complied with other relevant national regulations.

1.3.3.1 The Amending Directive

The UK DTI has argued to the Commission that since the required number of harmonized standards will take time to be drawn up and published, there should be some relaxation in the timetable for full implementation of the Directive. The argument is that as long as appropriate harmonized standards are not available, discrepancies in the interpretation of the essential requirements (as used within technical construction files) will lead to frequent disagreements as to whether the requirements are met. This in turn will diminish the efficiency of application of the Directive. The result of this argument has been for the Commission to propose to the Council of Ministers an Amending Directive.

The effect of the Amending Directive is to require member states to continue to permit the placing on the market or putting into service of apparatus which conforms to the national regulations in force in their territories before 31st December 1991, for a period of four years from this date. Therefore, it will not become legally essential to comply with the EMC Directive until 1st January 1996. The actual date is not absolutely fixed at the time of writing, because the Amending Directive must first be agreed by the Council of Ministers which will not happen until some time in 1992, but the length of the amended transition period is almost certain to be four years.

1.3.3.2 The transition period and product marketing

The virtue of the extended transition period for manufacturers is that it allows a more gradual phasing of EMC measures into new designs. As the second part of this book is meant to show, you can only implement EMC properly and efficiently if you design it in to a product from the start. Products already on the market in 1992 would have been designed years earlier, before the EMC requirements were generally appreciated, and in many if not most cases would need expensive, time consuming and inefficient measures to bring them into compliance. The extended transition period allows firms to tailor their marketing strategies to continue to market existing products without modification (and without the CE mark) until the end of 1995, whilst also designing measures into new products that will be launched nearer these dates to ensure that they will comply.

A typical design and development cycle of 2 years means that EMC should be an issue *immediately* for design teams. It would be entirely inappropriate for product managers to use the transition period as an excuse to do nothing about EMC for another two or three years.

1.3.4 Telecommunications terminal equipment

Apparatus defined as telecomms terminal equipment (TTE) has to comply with the requirements of a separate Directive (91/263/EEC) as well as those of the EMC Directive. This provides general telecomms requirements, as well as requirements for safeguarding the public telecommunications network, the safety of users and personnel and EMC requirements which are specific to TTE. Additionally article 10.5 of the EMC Directive requires radio transmitters (which may also be TTE, such as cellphone transmitters) to be type approved, which needs certification from a notified body.

Compliance with the TTE Directive's EMC requirements is on the basis of Common Technical Requirements (CTRs), which have yet to be written. The European Telecommunications Standards Institute (ETSI) is preparing to clarify the relationship between a European Telecomms Standard (ETS) and a CTR. It is as yet unclear to what extent the CTRs' scope will cover EMC and hence how far the EMC Directive itself will apply to TTE.

1.3.5 Testing and the competent body

Except in the case of products which it is clear will intrinsically not cause interference or be susceptible to it, such as the electric fire or pocket torch mentioned earlier, each manufacturer will need to submit products to some degree of EMC testing to be sure that they comply with the Directive. Chapter 3 considers EMC test methods in detail. To cover the eventual requirements of the standards, the scope of the tests will need to include mains harmonic, conducted and radiated RF emissions, plus immunity to RF, transients, electrostatic discharge and supply disturbances. A test facility to address all these phenomena at compliance level is beyond the budget of all but the largest companies. Not only are a screened room, an open area test site plus all the test equipment needed, but also the staff to run the facility – which itself requires a level of skill, experience and competence not usually found in most development or test departments. A large company may have the product volume and available capital which justifies investment (of the order of £1m) in an in-house facility of this nature. There are several such companies throughout Europe who have already taken this step. If they will be certifying exclusively to harmonized standards then no external constraints are placed on the operation of these in house test facilities. If they require competent body status in order to use the technical file route, then this is also possible provided that they have been accredited (see later).

1.3.5.1 Options for testing

Small to medium sized enterprises will not be able to afford their own full-scale test facilities and their choices are limited:

- join and help finance a consortium of similar companies which operates a test facility jointly for the benefit of its members;
- use an independent test house for all their EMC test requirements;
- establish a rudimentary EMC test capability in-house for confidence checking, and use an independent test house for compliance testing only.

The first option has not been established on a widespread basis in the UK yet, although there are precedents in the form of co-operative "research clubs" in other fields. The second option will be expensive and has the disadvantage that experience gained in testing your own products is not brought in house to apply to future products. The expense could be diluted by using cheaper, non-accredited test houses for confidence checking and saving the accredited test houses for full compliance testing. It is though more preferable to develop a close relationship with one test house with which you feel comfortable than to change test houses at will. And unfortunately the nature of EMC testing is that there are large measurement uncertainties to contend with, and there is no guarantee that a test at one facility will produce the same results as an apparently identical test at another. (This has given rise to the rather cynical strategy of hawking a marginal product around several test houses until a "pass" is achieved, on the basis that this is cheaper than optimizing the product design!)

1.3.5.2 In house testing

The problem of measurement uncertainty also applies to the third option, with possibly greater force because the confidence checks are done in a largely uncontrolled environment. Even for confidence checks, the equipment budget needed to carry them out is by no means negligible. It can be reduced by hiring expensive equipment at the appropriate time if the work load is light. A less obvious disadvantage is that not only must you invest in test equipment and facilities, but also in training staff to use them and in keeping up to date with the highly fluid world of EMC regulations and test methods. An external test house will have up to date equipment, facilities and expertise.

The advantage of the in house approach is that you can carry out testing at any stage of the product design and production cycle, and the process of EMC confidence testing helps to instil in the design team an awareness not only of the test techniques, but also of the effectiveness of the various design measures that are taken to improve EMC. The benefit of this will be reaped in future designs. Also, designers will be under much less stress if they have the ability to test and re-test modifications made at the bench without a concern for the money that is being spent in the process.

If the product will be certified to harmonized standards then there is no need to use an external test house at all, provided that you are confident in the capability and accuracy of your own tests. Nevertheless many firms, and especially their empowered signatory who signs the declaration of conformity, would be happier having independent confirmation of compliance from an organization whose competence in the field is recognized – and this is sometimes a commercial requirement anyway. It would be perfectly in order to choose some tests, perhaps those involving RF emissions or immunity, to be performed outside while others such as transient, ESD and mains disturbance immunity are done in-house.

The Atkins Report

The UK Department of Trade and Industry commissioned a study in 1989 to compare the demand for EMC testing and consultancy services, generated by the EMC Directive, with the capacity to supply such services provided by the existing test houses. This was subsequently published [124] and its most important conclusions were

- that the demand would exceed capacity to supply in the UK by a factor of about 6 at start-up, and about 2 in subsequent years;
- as a result a large backlog of work would accumulate, affecting

manufacturers' product launches;

- investment would be urgently needed in test facilities and in training EMC engineers. The key limiting factor on expansion of capacity would be the availability of personnel.

The methodology of the Atkins report came in for some criticism and the figure of a factor of 6 shortfall in supply was disputed; some commentators thought it would be higher than this, some less. Nevertheless the broad conclusions were accepted and several new test houses were inaugurated on the strength of the report. In fact, at the time of writing the anticipated flood of work for test houses has not materialized. This can be partly attributed to uncertainty among manufacturers as to how the EMC Directive applies to them, particularly with the confusion over the transition period and the delay in publishing standards, and partly to recessionary pressures which mean that most manufacturers have other more urgent things to worry about than complying with EC Directives.

If the conclusions of the Atkins report are even partly justified, when the Directive is finally active and fully operational the pressure on test house and consultancy resources is likely to increase dramatically.

1.3.5.3 Competent bodies

If the technical file route is chosen then you have to involve an independent competent body. The Directive lays down a number of requirements that must be met by anyone seeking competent body status:

- availability of personnel and of the necessary means and equipment;
- technical competence and professional integrity;
- independence of staff and technical personnel in relation to the product in question;
- maintenance of professional secrecy;
- possession of civil liability insurance.

Accreditation

Many of these requirements are met by accreditation, which in Europe is based on the EN45000 series of standards. This covers organisation and management, calibration and maintenance of test equipment, measurement traceability and procedures, records and reports, the quality system, and staff competence. In the UK the body which handles accreditation is NAMAS, the National Measurement Accreditation Service, based at the National Physical Laboratory. Mutual recognition of test house accreditation throughout Europe has yet to be achieved, and this is a major aim of the European Organisation for Testing and Certification (EOTC). The European groups responsible for accreditation of test facilities are given in Table 1.2.

Accreditation is a major requirement for appointment as a competent body for the purposes of the EMC Directive, but not the only one. In the UK, the Secretary of State for Trade and Industry will actually appoint competent bodies, and the DTI has indicated that a further requirement is the capability to make engineering judgements on the contents of a technical file, which is not a feature of test accreditation. At the same time, competent bodies will need to have adequate test facilities; independent consultants do not qualify.

A competent body must be resident within the EC. US, Japanese or other test houses outside the EC cannot apply for competent body status. It may be possible for a manufacturer to gain competent body status for his own test facility, assuming it meets the accreditation criteria, provided that it can demonstrate managerial independence from the groups responsible for the products being tested.

Austria	BMWA
Belgium	-
Denmark	STP
Finland	FINLAS
France	RNE
Germany	DAR/TGA/BAM
Greece	ELOT
Ireland	ILAB
Italy	SINAL
The Netherlands	STERLAB
Portugal	IPQ
Spain	RELE
Sweden	SWEDAC
Switzerland	OFM
UK	NAMAS

Table 1.2 European organisations responsible for test accreditation

1.3.6 Standards

The self-certification route (section 1.3.1) is the preferred route to demonstrating compliance with the Directive. This route depends on the availability of standards which can be applied to the product in question. The detail of the appropriate standards is covered in chapter 2; this section will discuss their general availability and applicability.

Prior to the adoption of the EMC Directive, the EMC standards regime had developed in a somewhat piecemeal fashion. The existing standards fell into one of a number of categories:

- RFI: intended to protect the radio spectrum from specific interference sources, such as information technology equipment, motor vehicle ignition, household appliances or fluorescent lights
- mains emissions: specifically harmonic currents and short-term variations, to protect the low-voltage power distribution network
- product- and industry-specific: to ensure the immunity from interference of particular types of product, such as process instrumentation or legal metrology, or to regulate emissions from equipment that will be used in a specific environment, such as marine equipment

These standards are not over-ridden by the Directive; those which have been harmonized by CENELEC may be applied to products within their scope and are regarded as adequate to demonstrate compliance. The same applies to non-harmonized standards which have been notified to and agreed by the Commission.

1.3.6.1 The generic standards

There are many industry sectors for which no product-specific standards have been developed. This is especially so for immunity; there are many more emission standards than there are for immunity. In order to fill this gap wherever possible, CENELEC have given a high priority to developing the Generic Standards. These are standards with a wide application, not related to any particular product or product family, and are intended to represent the essential requirements of the Directive. They are divided into two standards, one for immunity and one for emissions, each of which has separate parts for different environment classes (Table 1.3).

	Part 1	Part 2	Part 3
EN 50 081 Emissions EN 50 082 Immunity	Domestic, Commercial, Light Industrial	Industrial	Special

Table 1.3 The generic standards

Where a relevant product-specific standard does exist, this takes precedence over the generic standard. It will be common, though, for a particular product – such as a domestic appliance – to be covered by one product standard for mains harmonic emissions, another for RF emissions and the generic standard for immunity. All these standards must be satisfied before compliance with the Directive can be claimed. Other mixed combinations will occur until a comprehensive range of product standards has been developed – a process which will take several years.

Environment classes

The distinction between environmental classes is based on the electromagnetic conditions that obtain in general throughout the specified environments [78]. The inclusion of the "light industrial" environment (workshops, laboratories and service centres) in class 1 has been the subject of some controversy, but studies have shown that there is no significant difference between the electromagnetic conditions at residential, commercial and light industrial locations. Equipment for the class 2 "industrial" environment is considered to be connected to a dedicated transformer or special power source, in contrast to the class 1 environment which is considered to be supplied from the public mains network. At the time of writing the "special" environment has not been defined.

Referenced standards

The tests defined in the generic standards are based only on internationally approved, already existing standards. For each electromagnetic phenomenon a test procedure given by such a standard is referenced, and a single test level or limit is laid down. No new tests are defined in the body of any generic standard. Since the referenced standards are undergoing revision to incorporate new tests, these are noted in an "informative annex" in each generic standard. The purpose of this is to warn users of those requirements that will become mandatory in the future, when a new standard or the revision to the referenced standard is agreed and published.

There is some concern over whether a product which meets the initial version of the generic standards will have to be re-certified whenever a new version, with an extended list of required tests, is published. CENELEC rules allow a period of grace between the publication of a new revision and the withdrawal of the old revision, during which certification to either is valid. In any case, it is the intention of TC110 (the committee which oversees the generic standards) to ensure that there is a period of consolidation stretching into some years, to gain experience with applying the generic standards, before a new revision is published. Thus the informative annex will remain informative only for some time to come.

1.3.6.2 Performance criteria

A particular problem with immunity is that the equipment under test may exhibit a wide variety of responses to the test stimulus. This can range from a complete lack of response, through a degradation in the accuracy of measured variables to total corruption of its operation. The same problem does not exist for emissions, where comparison with a defined test limit is possible. To account for this variety, the generic immunity standards include three generalised performance criteria for the purpose of evaluating test results. In the test report, you must include a functional description and a specific definition of performance criteria based on these, during or as a consequence of the EMC testing. The definitions of these criteria are as follows:

- Performance criterion A (continuous phenomena): The apparatus shall continue to operate as intended. No degradation of performance or loss of function is allowed below a performance level specified by the manufacturer, when the apparatus is used as intended. In some cases the performance level may be replaced by a permissible loss of performance. If the minimum performance level or the permissible performance loss is not specified by the manufacturer then either of these may be derived from the product description and documentation (including leaflets and advertising) and what the user may reasonably expect from the apparatus if used as intended.
- Performance criterion B (transient phenomena): The apparatus shall continue to operate as intended after the test. No degradation of performance or loss of function is allowed below a performance level specified by the manufacturer, when the apparatus is used as intended. During the test, degradation of performance is however allowed. No change of actual operating state or stored data is allowed. If the minimum performance level or the permissible performance loss is not specified by the manufacturer then either of these may be derived from the product description and documentation (including leaflets and advertising) and what the user may reasonably expect from the apparatus if used as intended.
- Performance criterion C (mains interruption): Temporary loss of function is allowed, provided the loss of function is self recoverable or can be restored by the operation of the controls.

1.3.6.3 Basic standards

Those standards which are referenced in the generic standards, for example the various parts of IEC 801 and IEC 1000, are known as "basic" standards. The exact definition of "basic" has yet to be agreed, but it can be regarded as meaning those standards that are entirely devoted to aspects of EMC that will prove to be of general interest and use to all committees developing other standards – for instance, product specific standards.

1.4 Action for compliance

The steps which you will need to take in order for a new product to achieve compliance with the EMC Directive and bear the CE mark can be summarized as follows (ignoring radio transmitters, which must be type approved, and telecomms terminal equipment, which are subject to their own Directive).

A. Self certification

1. From the marketing specification, determine what type of product it will be and what environment it will be sold for use within, and hence which if any product-specific standards (see section 2.2 and 2.3) apply to it. If your company only ever makes or imports products for one particular application then you will be able to use the same product-specific standard(s) for all products.
2. If no product-specific standards apply, check the generic standards to see if the tests specified in them are applicable. The environmental classification will depend on the intended power supply connection.
3. If you cannot apply the generic standards, or don't wish to for the reasons discussed in section 1.3.2, then you will need to follow the technical file route (B).
4. Having determined what standards you will use, decide on the test levels and to what ports of the equipment (enclosure, power leads, signal/control leads) they will apply. In some cases there will be no choice, but in others the test applicability will depend on factors such as length of cable, EUT configuration and class of environment.
5. From this information you will be able to draw up a test plan, which specifies in detail the version and configuration of the EUT and any associated apparatus, the tests that will be applied to it and the pass/fail criteria. Test plans are covered in greater depth in Appendix B. You can discuss this with your selected test house or your in-house test facility staff, and it will form the basis for your contract with them and also for the technical documentation required by the provisions of the Directive.
6. Knowing the requirements of the test plan will enable you to some degree to incorporate cost effective EMC measures into the product design, since the test limits and the points to which they will be applied will have been specified.
7. As the design progresses through prototype and pre-production stages you can make pre-compliance confidence tests to check the performance of the product and also the validity of the test plan. It is normal for both design and test plan to undergo iterative modifications during this stage.
8. Once the design has been finalised and shortly before the product launch you can then perform the full compliance tests on a production sample, the results of which are recorded in the technical documentation. Provided that confidence tests were satisfactory this should be no more than a formality.
9. You are then at liberty to mark the product, and/or its packaging or documentation with the CE mark (provided that there is no other Directive to satisfy) and your empowered signatory can sign the Declaration of

Conformity, to be kept for ten years. The product can be placed on the market.

10. Once the product is in series production you must take steps to ensure that it continues to comply with the protection requirements.

B. Technical construction file

1. Your product is such that you cannot or will not apply harmonized standards to it to cover the essential requirements in full. In this case, you must work with a competent body.
2. Having chosen a competent body (in the UK, the DTI maintains a list of these – it is also possible to work with a competent body in any other Member State), discuss with them the design features and test requirements that they would need to see to satisfy them that the product complied with the Directive. From this discussion you should be able to draw up a test plan as in A.5 above. From this point, you can continue as in A.6 and 7.
3. By the time the product has been finalised you will have created the technical construction file as described in section 1.3.2. This may or may not include compliance tests as agreed with, and possibly but not necessarily performed by, the competent body. The competent body will then review the complete technical file and, provided they are satisfied, will issue a report or certificate to say so.
4. At this point you are at liberty to proceed as in A.9 and 10 above.

Chapter 2

Standards

2.1 The standards making bodies

The structure of the bodies which are responsible for defining EMC standards for the purposes of the EMC Directive is shown in Figure 2.1.

2.1.1 The International Electrotechnical Commission

The IEC operates in close co-operation with the International Standards Organization (ISO) and in 1990 had 41 member countries. It is composed of National Committees which are expected to be fully representative of all electrotechnical interests in their respective countries. Work is carried out in technical committees and their sub-committees addressing particular product sectors, and the secretariat of each technical committee is the responsibility of one of the 41 National Committees, which appoints a Secretary with the necessary resources. The IEC's objectives are "to promote international co-operation on all questions of standardization.... (this is) achieved by issuing publications including recommendations in the form of international standards which the National Committees are expected to use for their work on national standards."[51]

Two IEC technical committees are devoted full time to EMC work, although nearly forty others have some involvement with EMC as part of their scope. The two full time committees are TC77, *Electromagnetic compatibility between equipment including networks*, and the *International Special Committee on Radio Interference* or CISPR, which is the acronym for its French title. Co-ordination of the IEC's work on EMC between the many committees involved is the responsibility of ACEC, the Advisory Committee on EMC, which is expected to ensure against the development of conflicting standards.

In future the major output of the technical committees will be as parts of IEC Publication 1000, *Electromagnetic Compatibility*. This document is being published in stages as defined by the plan shown in Table 2.1, and will incorporate all non-CISPR EMC material. Current standards such as IEC 555 and IEC 801 are expected to be eventually subsumed within IEC 1000. CISPR publications deal with limits and methods of measurement of the radio interference characteristics of potentially disturbing sources, and will probably continue to co-exist with IEC 1000. One further important document is Chapter 161 of IEC Publication 50 [98], the International Electrotechnical Vocabulary. This contains definitions of EMC terminology in English, French and Russian, with equivalent terms in Dutch, German, Italian, Polish, Spanish and Swedish.

IEC standards themselves have *no legal standing* with regard to the EMC Directive. If the National Committees do not agree with them, they need not adopt them; although in the UK, 85% of IEC standards are transposed to British Standards.

IEC1000-1	Part 1: General General considerations (introduction, fundamental principles) Definitions, terminology
IEC1000-2	Part 2: Environment Description of the environment Classification of the environment Compatibility levels
IEC1000-3	Part 3: Limits Emission limits Immunity limits (if not the responsibility of product committees)
IEC1000-4	Part 4: Testing and measurement techniques Measurement techniques Testing techniques
IEC1000-5	Part 5: Installation and mitigation guidelines Installation guidelines Mitigation methods and devices
IEC1000-9	Part 9: Miscellaneous

IEC1000 is being published in separate parts by IEC TC77 according to the above plan. Each part is further subdivided into sections which can be published either as international standards or as Technical Reports.

Table 2.1 Plan of IEC 1000 [51]

2.1.2 CENELEC

CENELEC (the European Organisation for Electrotechnical Standardisation) is the European standards making body, which has (among many other things) been mandated by the Commission of the EC to produce EMC standards for use with the European EMC Directive. For telecommunications equipment ETSI (the European Telecommunications Standards Institute) is the mandated standards body. ETSI generates standards for telecomms network equipment that is not available to the subscriber, and for radio communications equipment and broadcast transmitters.

CENELEC and ETSI use IEC/CISPR results wherever possible as a basis for preparation of drafts for such standards, and the committee charged with the duty of preparing the EMC standards is TC110. Representatives of National Committees meet in TC110 about once a year to discuss the technical implementation of the drafts. TC110 has a sub-committee, SC110A, which is concerned specifically with immunity of Information Technology Equipment (ITE), and three other working groups, one of which is responsible for the Generic Standards (section 1.3.6.1 on page 26).

CENELEC is made up of the National Committees of each of the EC and EFTA countries; adoption of standards is based on a qualified weighted voting by the 18 National Committees [26][83]. Of these member committees France, UK, Germany and Italy have 10 votes, Spain has 8 votes and the other countries have between 3 and 5 votes. Unlike the position with international standards, a country must accept a new CENELEC standard even if it voted against it. Formal national conditions may be attached to the standard to ameliorate this situation, such as the occasion when CENELEC decided to harmonise on a 230V mains supply, and the UK declared to stay at 240V as a special national condition.

In the UK the BSI committee GEL110 generates the British position on TC110 papers. The BSI has an obligation to invite all organizations which have an interest in EMC to be members of GEL110 – in practice this is done mostly through representation by trade organizations.

Once CENELEC has produced and agreed a European EMC standard (prefixed with EN or HD) all the CENELEC countries are required to implement identical national standards. The EN will be transposed word for word, while the HD (harmonisation document) does not need to be reproduced verbatim as long as it reflects the technical content. The reference number of the EN and the equivalent national standards will then be published in the Official Journal of the European Communities, and once this is done the standards are deemed to be "relevant standards" for the purpose of demonstrating compliance with the Directive. Conflicting national standards must be withdrawn within a limited time frame.

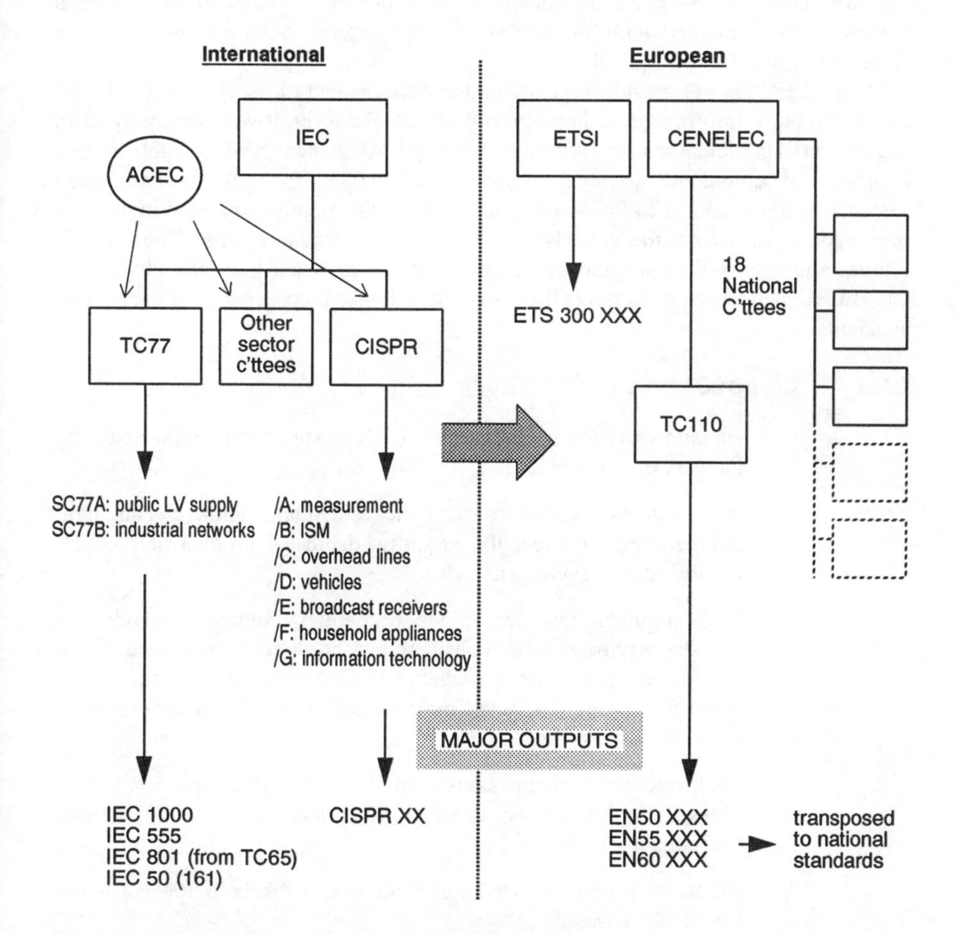

Figure 2.1 Important standards makers for the EMC Directive

Draft standards and amendments to existing standards are made available for public comment (through the National Committees) for some time before the standard is actually published. Apart from being the mechanism by which industry can influence the content of the standards, this has the further advantage of permitting manufacturers to make an informed decision on the testing and limit levels to which they may choose to submit their products in advance of the actual publication date, even though it is not possible to make an official declaration of compliance with an unpublished standard. There is of course some risk that the final published version will differ in detail, and sometimes quite substantially, from the draft.

2.2 Standards relating to the EMC Directive - emissions

The following section outlines those standards which are known at the time of writing to be either available or to have been made available for public comment (prefixed with *pr*) and which will form harmonized standards or basic standards for the purposes of the EMC Directive. Note that the contents of preliminary standards are by no means finalised. Those marked with an asterisk (*) have already been announced in the Official Journal of the EC [120].

CENELEC has put greatest urgency on the development of generic standards and on standards for information technology equipment. However, it will frequently be in the interest of particular industry sectors to develop EMC standards which apply to their own types of equipment (early examples are pr EN 50 083-2 for cable TV distribution systems, or pr ETS 300 126/EN 50 096 for ISDN interface equipment), and in the next few years we can expect to see a whole raft of such standards to emerge. The unstable interim situation will hasten the development of these. When they are adopted by CENELEC then these product-specific standards will take precedence over the generic standards.

2.2.1 EN 50 081 part 1

Title Generic emission standard, part 1: Domestic, commercial and light industry environment

Scope All apparatus intended for use in the residential, commercial and light industrial environment for which no dedicated product or product-family emission standards exist

NB equipment installed in the residential, commercial and light industry environment is considered to be directly connected to the public mains supply or to a dedicated DC source. Typical locations are residential properties, retail outlets, laboratories, business premises, outdoor locations etc.

Tests Enclosure: radiated emissions from 30 to 1000MHz as per EN 55 022 Class B; applicable only to apparatus containing processing devices operating above 9kHz

AC mains port: conducted emissions from 150kHz to 30MHz as per EN 55 022 Class B

Discontinuous interference on AC mains port measured at spot frequencies as per EN 55 014, if relevant

Mains harmonic emission measured as per EN 60 555 part 2 (note that application is limited by the scope of EN 60 555-2)

NB an informative annex references tests which will be proposed for inclusion in the standard when the relevant reference standards are published. This includes tests on signal, control and DC power ports: conducted current from 150kHz to 30MHz as per draft amendment to EN55022

2.2.2 pr EN 50 081 part 2

Title Generic emission standard, part 2: industrial environment

Scope Apparatus operating at less than 1000V_{rms} AC intended for use in the industrial environment, for which no dedicated product or product-family immunity standard exists, but excluding radio transmitters.

NB equipment installed in the industrial environment is not connected to the public mains network but is considered to be connected to an industrial power distribution network with a dedicated distribution transformer.

Tests Enclosure: radiated emissions from 30 to 1000MHz as per EN55011

AC mains port: conducted emissions from 150kHz to 30MHz as per EN55011; impulse noise appearing more often than 5 times per minute must comply with EN55014

NB an informative annex references tests which will be proposed for inclusion in the standard when the relevant reference standards are published. This includes tests on signal, control and DC power ports: conducted emissions from 150kHz to 30MHz as per draft amendment to EN55022

2.2.3 EN 55 011 *

Title Limits and methods of measurement of radio disturbance characteristics of industrial, scientific and medical (ISM) radio-frequency equipment

Equivalents CISPR11, BS4809 (not direct equivalent, to be withdrawn)

Scope Equipment designed to generate RF energy for industrial, scientific and medical (ISM) purposes, including spark erosion

Class A equipment is for use in all establishments other than domestic; Class B equipment is suitable for use in domestic establishments.

Group 1 equipment is that in which the RF energy generated is necessary for its internal functioning; Group 2 equipment is that in which RF energy is generated for material treatment and spark erosion.

Tests Mains terminal disturbance voltage from 150kHz to 30MHz measured on a test site using 50Ω/50μH CISPR artificial mains network; Group 2 Class A equipment subject to less stringent limits

Radiated emissions from 30MHz to 1000MHz on a test site (Class A or B) or in situ (Class A only); Group 2 Class A equipment to be measured from 0.15 to 1000MHz but with relaxed limits, below 30MHz measurement performed with loop antenna

NB limits for Group 1 and Group 2 Class B equipment from 0.15 to 30MHz are under consideration

2.2.4 EN 55 013 *

Title Limits and methods of measurement of radio disturbance characteristics of broadcast receivers and associated equipment

Equivalents CISPR13, BS905 part 1, VDE0872 part 13

Scope Broadcast sound and television receivers, and associated equipment intended to be connected directly to these or to generate or reproduce audio or visual information, for example audio equipment, video cassette recorders, compact disc players, electronic organs. Information technology equipment as defined in EN55022 is excluded

NB EN55013:1990 removes audio equipment from the scope of EN55014

Tests Mains terminal interference voltage from 150kHz to 30MHz measured using 50Ω/50μH CISPR artificial mains network.

Antenna terminal disturbance voltage over the range 30 to 1000MHz due to local oscillator and other sources, less stringent limits for car radios

Radiated disturbance field strength of local oscillator and harmonics in the range 80 to 1000MHz measured on an open area test site at a distance of 3m

Disturbance power of associated equipment excluding video recorders on all leads of length 25cm or more, over the range 30 to 300MHz, measured by means of the absorbing clamp

Limits Limits for mains terminal disturbance voltage and disturbance power are the same as those in EN55014. Radiated field limits for local osciullator and harmonics are 12 - 20dB higher than equivalent Class B emissions limits for other products

2.2.5 EN 55 014 *

Title Limits and methods of measurement of radio interference characteristics of household electrical appliances, portable tools and similar electrical apparatus

Equivalents CISPR14, BS800, VDE0875 part 1

Scope Electrical household appliances, portable tools under 2kW and other electrical equipment causing similar types of continuous or discontinuous interference, including regulating controls under 25A incorporating semiconductor devices.

NB there is a revision pending which excludes apparatus covered by other standards, and also excludes power supplies to be used separately. This revision makes major changes to other parts of EN55014

Tests Mains terminal disturbance voltage from 150kHz (revision to 148.5kHz) to 30MHz measured using 50Ω/50μH CISPR artificial mains network; less stringent limits for portable tools and the load terminals of regulating controls. Draft revision will require average as well as quasi peak measurement. Discontinuous interference (clicks) must also be measured at spot frequencies for appliances which generate such interference through switching operations

Interference power from 30MHz to 300MHz on mains lead, measured by means of the absorbing clamp; regulating controls incorporating semiconductor devices excluded

2.2.6 EN 55 015 *

Title Limits and methods of measurement of radio interference characteristics of fluorescent lamps and luminaires

Equivalents CISPR15, BS5394, VDE0875 part 2

Scope Conduction and radiation of electromagnetic energy from fluorescent lamps and luminaires

Tests For luminaires with replaceable lamps, insertion loss is measured at 5 frequencies between 160kHz and 1400kHz between terminals on a dummy lamp (construction specified in the standard) and the mains terminals of the luminaire

NB for self-ballasted fluorescent lamps, or when it is impossible to replace the lamp with a dummy lamp or when an electronic starter is incorporated, the limits of EN55014 in the frequency range 150kHz to 30MHz apply to tests made using the 50Ω/50μH CISPR artificial mains network

2.2.7 EN 55 022 *

Title Limits and methods of measurement of radio interference characteristics of information technology equipment

Equivalents CISPR22, BS6527, VDE0878 part 3

Scope Equipment designed for the purpose of receiving data from an external source, performing some processing functions on the received data and providing a data output.

NB there is a draft amendment pending which would replace this definition by equipment whose primary function is either (or a combination of) data entry, storage, display, retrieval, transmission, processing, switching or control, and which may be equippped with one or more terminal ports typically operated for information transfer, and with a rated supply voltage not exceeding 600V.

Class A equipment is for use in typical commercial establishments; Class B equipment is suitable for use in domestic establishments

Tests Mains terminal interference voltage, quasi-peak and average detection from 150kHz to 30MHz measured using 50Ω/50μH CISPR artificial mains network.

Radiated interference field strength using quasi peak detection from 30MHz to 1000MHz measured at 10m on an open area test site

NB there is a draft amendment pending which will require measurements also on telecommunications signal ports, of conducted current over the range 150kHz to 30MHz. Various other amendments to do with operating conditions, test set-up, site attenuation and testing of physically large systems etc. are under consideration. A complete revision of CISPR 22, with a consequent revision of EN 55 022, is likely to be proposed during 1992.

2.2.8 EN 60 555 parts 2 and 3 *

Title Disturbances in supply systems caused by household appliances and similar electrical equipment, part 2 : Harmonics, part 3: voltage fluctuations

Equivalents IEC 555, BS 5406, VDE0838

Scope Electrical and electronic equipment for household and similar use, intended to be connected to low-voltage ac distribution systems (240V single phase or 415V three-phase)

NB there is an amendment pending which would widen the scope to all electrical and electronic equipment having an input current up to 16A and intended to be connected to public low-voltage distribution systems

Tests Part 2: measurement of 50Hz harmonic currents up to 2kHz using a wave analyser and current shunt or transformer

Part 3: measurement of voltage fluctuations using a flickermeter or by analytical methods

Limits Part 2: present version has absolute limits in amps applying to each harmonic from n = 2 to 40. Proposed amendment divides equipment into four classes depending on its type and power rating and applies different limits to each. This amendment is still the subject of much controversy

Part 3: limits apply to magnitude of maximum permissible percentage voltage changes with respect to number of voltage changes per second or per minute

2.3 Standards relating to the EMC Directive - immunity

2.3.1 EN 50 082 part 1

Title Generic immunity standard, part 1: domestic, commercial and light industry environment

Scope All apparatus intended for use in the residential, commercial and light industrial environment for which no dedicated product or product-family immunity standards exist

Tests Electrostatic discharge to enclosure as per IEC 801 part 2, at 8kV (air discharge)

Radiated RF field from 27MHz to 500MHz as per IEC 801 part 3, at 3V/m

Electrical fast transients 5/50ns common mode as per IEC 801 part 4, applied to all I/O and power ports with some exceptions, amplitude 0.5 or 1kV dependent on type of port and method of coupling

NB an informative annex references tests which will be proposed for inclusion in the standard when the relevant reference standards are published. This includes:

- magnetic field, 50Hz at 3A/m
- extension of RF radiated field to 80 – 1000MHz AM
- pulsed RF field, 1.89GHz at 3V/m
- inclusion of contact disharge ESD at 4kV
- conducted RF common mode voltage on all I/O and power ports, 150kHz to 100MHz at 3V, as per draft IEC 801 part 6
- AC 50Hz common mode voltage of 10V on signal and control lines
- supply voltage deviations, interruptions and fluctuations on supply ports
- surges on AC power ports, 1kV differential mode, 2kV common mode as per draft IEC 801 part 5

NB the applicability of many of the above tests depends on the allowable length of line that may be connected to the port in question

Criteria Three performance criteria for test results are proposed:

- the apparatus continues to operate as intended with no degradation below a performance level specified by the manufacturer;
- the apparatus continues to operate as intended after the test, but during the test some degradation of performance is allowed;
- temporary loss of function is allowed, provided that it is self- or operator-recoverable

2.3.2 pr EN 50 082 part 2

Title Generic immunity standard, part 2: industrial environment

Scope Apparatus operating at less than 1000V_{rms} AC intended for use in the industrial environment, for which no dedicated product or product-family immunity standard exists, but excluding radio transmitters

Tests Electrostatic discharge to enclosure as per IEC 801 part 2, at 8kV (air discharge) or 4kV (contact discharge)

Radiated RF field from 27MHz to 500MHz as per IEC 801 part 3, at 10V/m

Electrical fast transients 5/50ns common mode as per IEC 801 part 4, applied to all I/O and power ports, amplitude 1 or 2kV dependent on type of port and method of coupling

NB an informative annex references tests which will be proposed for inclusion in the standard when the relevant reference standards are published. This includes:

- magnetic field, 50Hz at 30A/m
- extension of RF radiated field to 26 - 1000MHz
- pulsed RF field, 1.89GHz at 3V/m
- conducted RF common mode voltage on all I/O and power ports, 150kHz to 100MHz at 3V or 10V depending on type of port, as per draft IEC 801 part 6
- AC 50Hz common mode voltage of 10V or 20Vrms depending on type of port, on signal and control lines
- supply voltage deviations, interruptions and fluctuations on supply ports
- Surges on AC power ports, 2kV differential mode, 4kV common mode as per draft IEC 801 part 5

NB the applicability of many of the above tests depends on the allowable length of line that may be connected to the port in question

Criteria Three performance criteria for test results are proposed:

- the apparatus continues to operate as intended with no degradation below a performance level specified by the manufacturer;
- the apparatus continues to operate as intended after the test, but during the test some degradation of performance is allowed;
- temporary loss of function is allowed, provided that it is self- or operator-recoverable

2.3.3 pr EN 50 093

Title Basic immunity standard for voltage dips, short interruptions and voltage variations

Scope Electrical and electronic equipment fed by low voltage power supply networks and having an input current not exceeding 16A per phase, but not equipment which is connected to DC networks or 400Hz AC networks

Tests Dips and short interruptions initiated at the zero crossing of the input voltage, to a level of 0%, 40% and 70% of the nominal voltage for a duration of 0.5 to 50 periods

Short term variations to a level of 40% and 0% of nominal voltage, taking 2.0 seconds to reach the test level and to recover from it, and 0.5 second at the test level

Long term deviations with a test duration sufficient to allow the operating temperature to stabilise, at +12% and −15% of nominal voltage

Criteria Test results to be classified as follows:

- normal performance within specification limits
- temporary degradation or loss of function or performance which is self recoverable
- temporary degradation or loss of function or performance which requires operator intervention or system reset
- degradation or loss of function which is not recoverable due to hardware or software damage or loss of data

2.3.4 EN 55 020 *

Title Immunity from radio interference of broadcast receivers and associated equipment

Equivalents CISPR 20, BS 905 part 2, VDE0872 part 20

Scope Broadcast sound and television receivers, and associated equipment intended to be connected directly to these or to generate or reproduce audio or visual information, for example audio equipment, video cassette recorders, compact disc players, electronic organs. Information technology equipment as defined in EN 55 022 is excluded

No immunity requirements apply to battery powered sound and tv receivers or those without an external antenna connection

Tests Immunity from unwanted signals present at the antenna terminal: VHF band II receivers tested with in-band and out-of-band signals up to 85dBμV; tv receivers tested with adjacent channel modulated signals up to 80dBμV

Immunity from conducted voltages at the mains input, audio input and output terminals of receivers (except AM sound and car radios) and multi-function equipment over the range 150kHz to 150MHz; audio input & output terminals have less stringent low frequency levels than mains, loudspeaker and headphone terminals; the tuned channel and

IF channel frequencies are excluded

Immunity from conducted currents of receivers (including car radios and AM sound) and multi-function equipment over the range 26 to 30MHz applied to the antenna terminal

Immunity from radiated fields from 150kHz to 150MHz of receivers and multi-function equipment, as tested in an open stripline test set up, at 125dBμV/m except at IF and in-band frequencies

NB Limits of immunity for associated equipment in all cases are under consideration.

Criteria Wanted to unwanted audio signal ratio of ≥40dB, or just perceptible degradation of a standard picture

2.3.5 pr EN 55 101

Title Immunity requirements for Information Technology Equipment

Equivalents CISPR 24 (draft) (note that pr EN 55 101 will be renumbered EN 55 024 to align with the CISPR numbering)

Scope Information Technology Equipment as defined in EN 55 022.

Tests Part 2 (draft): Electrostatic Discharge. 200 contact discharges each at negative and positive polarity to at least 4 test points, one of which must receive 50 discharges in indirect contact. Discharges applied only to points and surfaces which are accessible to personnel during normal usage, including customer maintenance, but not to contacts of open connectors. Where contact discharges are not possible ten single air discharges applied to selected user accessible test points where breakdown may occur.

Test levels are 4kV for contact and 8kV for air discharge.

Part 3 (draft): Radiated RF disturbances. Radiated RF field in an absorber lined chamber or TEM cell at a strength of 3V/m amplitude modulated 80% at 1kHz over a frequency range of 80MHz to 1000MHz.

Part 4 (draft): Continuous conducted RF interference. RF current at 86dBμA (level established into a 50Ω calibration jig) and amplitude modulated 80% at 1kHz over the frequency range 150kHz to 80MHz, applied to all telecommunication, data, signal and power ports sequentially. Ports which are not intended to connect to data cables longer than 10m need not be tested below 30MHz. (Note that part 4 is less well advanced in drafting in comparison to parts 2 and 3, and will probably be re-numbered to correspond with IEC 801 part 6 when eventually published)

Criteria The EUT shall withstand the applied tests without damage, and shall operate correctly within its specified limits. Corruption of software or data associated with the EUT is not permitted. Temporary disturbances which are corrected, or are automatically or manually cleared immediately, are acceptable (ESD testing); momentary degradation may be considered acceptable provided the malfunction criteria are documented in the test report (radiated RF testing).

2.3.6 IEC 801

Title Electromagnetic compatibility for industrial-process measurement and control equipment

Equivalents HD 481, BS 6667, VDE 0843

Scope The susceptibility of industrial-process measurement and control equipment to:

Part 2: electrostatic discharge generated by operators and between objects in close proximity

Part 3: radiated electromagnetic energy

Part 4: repetitive electrical fast transients

Part 5 (draft): surges caused by overvoltages/currents from switching and lightning transients

Part 6 (draft): conducted radio frequency disturbances

NB although originally written to apply to process measurement and control equipment, IEC 801 has been identified as a basic EMC document (to be re-published as part of IEC 1000) and therefore the test methods etc. to be found in it are intended to apply to all electrical and electronic equipment.

Part 2

Tests At least ten single discharges to preselected points, accessible to personnel during normal useage, in the most sensitive polarity. Contact discharge method to be used unless this is impossible, in which case air discharge used. Also ten single discharges to be applied to a coupling plane spaced 0.1m from the EUT.

Limits Severity levels from 2kV to 15kV (8kV contact discharge) depending on installation and environmental conditions.

Part 3

Tests Radiated RF field generated by antennas in a shielded anechoic enclosure, or by a stripline or TEM cell, swept from 27MHz to 500MHz at 1.5 . 10^{-3} decades/s with the EUT in its most sensitive orientation.

NB there is a draft amendment pending which will increase the frequency range to 26MHz – 1000MHz and require the signal to be amplitude modulated.

Limits Severity levels of 1, 3 or 10V/m (or greater) depending on the expected EMR environment.

Part 4

Tests Bursts of 5ns/50ns pulses at a repetition rate of 5kHz with a duration of 15ms and period of 300ms, applied in both polarities between power supply terminals (including the protective earth) and a reference ground plane, or via a capacitive coupling clamp onto I/O circuits and communication lines.

Limits Severity levels of 0.5, 1, 2 and 4kV on power supply lines, and half these values on signal, data and control lines, depending on the expected environmental and installation conditions.
NB an amendment to IEC 801 part 4 is likely to be proposed in 1993

Part 5 (draft)

Tests At least 5 positive and 5 negative surges of 1.2/50μs voltage or 8/20μs current waveshape surges from a surge generator of 2Ω output impedance, line to line on ac/dc power lines; 12Ω output impedance, line to earth on ac/dc power lines; 42Ω output impedance, capacitively coupled line to line and line to earth on I/O lines.

Limits Severity levels of 0.5, 1, 2 and 4kV, depending on environment and type of installation
NB IEC 801 part 5 is likely to be adopted in 1992

Part 6 (draft)

Tests RF voltage amplitude modulated 80% at 1kHz swept at $1.5 \cdot 10^{-3}$ decades/s over the frequency range 150kHz to 26MHz and 26MHz to 230MHz, applied via coupling/decoupling networks to each cable port of the EUT

NB: applicability of tests over the frequency range 26MHz to 230MHz overlaps with IEC 801 part 3 and may be used instead of the tests specified in that document, depending on the EUT dimensions

Limits Severity levels of 1, 3 or 10V emf depending on the EMR environment on final installation

2.4 Other standards

2.4.1 FCC Rules

In the US, radio frequency interference requirements are controlled by the FCC (Federal Communications Commission), which is an independent government agency responsible for regulating inter-state and international communications by radio, television, satellite and cable. The requirements are detailed in CFR (Code of Federal Regulations) 47. Part 15 of these regulations until 1990 applied to restricted and incidental radiation devices, that is those devices which emit RF interference as a by-product of their operation, or at very low power. In 1990 the regulations were revised into a new format which distinguishes primarily between unintentional and intentional radiators, and verifications or applications for certification made on or after June 23rd

1992 must comply with the new revision.

Part 15 subpart B, applying to unintentional radiators, includes clauses which cover specific classes of device such as power line carrier systems, TV receivers and TV interface devices. Industrial, scientific and medical devices which intentionally generate RF energy are covered under Part 18 of the rules. But the major impact of Part 15 is on those products which incorporate digital devices.

Digital devices

A "digital device" (previously defined as a computing device) is any electronic device or system that generates and uses timing signals or pulses exceeding 9kHz and uses digital techniques. Two classes are defined, depending on the intended market: class A for business, commercial or industrial use, and class B for residential use. These classes are subject to different limits, class B being the stricter. Before being able to market his equipment in the US, a manufacturer must either obtain certification from the FCC if it is a personal computer or associated peripheral, or must verify that the device complies with the applicable limits. Certification requires FCC approval before marketing is allowed, while verification merely requires that the manufacturer satisfies himself that the equipment meets the technical requirements.

There are some quite broad exemptions from the rules depending on application. These include digital devices used in transport vehicles, industrial plant or public utility control systems, industrial, commercial and medical test equipment, specialized medical computing equipment and a digital device used in an appliance.

Limits apply to conducted interference on the mains lead between 450kHz and 30MHz, and radiated interference measured either at 10m or 3m from 30MHz to 960MHz and above. The limits are similar but not identical to those laid down in CISPR-derived standards (see section 2.2). A major effect of the new revision is to extend the upper frequency limit to a possible maximum of 40GHz, depending on the frequencies used within the device. The relationship between internal clock (or other) frequencies and the maximum measurement frequency is shown in Table 2.2. From this you can see that devices with clock frequencies exceeding 108MHz must be tested for emissions well into the microwave region, a requirement which will be new to many test houses.

Highest frequency generated or used in the device or on which the device operates or tunes (MHz)	Upper frequency of measurement range (MHz)
Below 1.705	30
1.705 – 108	1000
108 – 500	2000
500 – 1000	5000
Above 1000	5th harmonic of highest frequency or 40GHz, whichever is lower

Table 2.2 Maximum measurement frequency for digital devices, FCC Rules Part 15 (1990)

The measurement techniques and defined test configurations differ from the CISPR standards; these are defined in FCC publication MP-4 [126] for computing devices, and other MP publications for more specific types of equipment. The requirements of MP-4 are discussed in more detail in section 3.1.4.2.

The American standard for measuring instrumentation and test methods, roughly equivalent in some areas to CISPR Publication 16, is ANSI C63.

2.4.2 German standards

In Germany, equipment operating at frequencies above 10kHz may not be used without a licence. A "General Licence" is granted to equipment which meets class B limits of the applicable standard. Alternatively a type-specific licence can be obtained for equipment which meets the class A (less stringent) limits. The user must ensure that the equipment they intend to use is covered by a licence, and so all manufacturers of equipment for the German market need to submit samples for type testing to the German VDE (Verband Deutscher Elektrotechniker) or an equivalent approved organisation. The applicable standards are VDE 0875 for broadband interference (as generated by household appliances, etc.) and VDE 0871 for broad- and narrow-band interference (as generated by information technology equipment for example). These standards are very broadly equivalent to EN 55 014 and EN 55 022 respectively. A notable exception is that VDE 0871 class B regulates conducted emissions below 150kHz down to 9kHz, whereas EN 55 022 has no requirement below 150kHz, also it requires radiated magnetic field tests below 30MHz, a requirement which is missing from EN 55 022.

With the implementation of the EMC Directive, the requirement to comply with VDE standards will be superseded by compliance with the standards nominated by the Directive and demonstrated by the presence of the CE mark. Meanwhile, during the transition period of the Directive (section 1.3.3) if a CE mark is not affixed the VDE standards and licensing laws will continue to apply to equipment marketed within Germany.

2.4.3 Other non-harmonized standards

The BSI publishes some other standards which are not for the present expected to be harmonized with European standards but which refer to electromagnetic compatibility. These are:

BS1597	Limits and methods of measurement of electromagnetic interference generated by marine equipment and installations
BS5049	Methods for measurement of radio interference characteristics of overhead power lines and high voltage equipment
BS5602	Code of practice for abatement of radio interference from overhead power lines
BS6345	Method for measurement of radio interference terminal voltage of lighting equipment
BS7027	Limits and methods of measurement of immunity of marine electrical and electronic equipment to conducted and radiated electromagnetic interference
3G100 : Part 4 : Section 2	General requirements for equipment for use in aircraft: electromagnetic interference at radio and audio frequencies

AU 243 Methods of test for electrical disturbance by conduction and coupling (automobile series, equivalent to ISO 7637)

There are also various standards for RF and transient immunity, in addition to those discussed earlier, which are written specifically for certain products or product sectors, or which are developed by large customers for in house or contractual use. These are generally comparable in scope and limit levels to IEC 801.

2.4.4 RF emissions limits

Most of the standards within the EN 55 0XX series have harmonised limit levels for conducted and radiated emissions. These standards derive from CISPR and the limit levels are set in each case for the same purpose, to safeguard the radio spectrum for other users. A minimum separation distance is assumed between source and susceptible equipment for this purpose.

Figure 2.2 and Figure 2.3 show the limits in graphical form for the emissions standards discussed above. FCC and VDE levels differ somewhat from the harmonised EN levels and are included for comparison. All radiated emission levels are normalised to a measuring distance of 10m. You should also bear in mind that there are detailed differences in the measurement methods between FCC, VDE and CISPR standards.

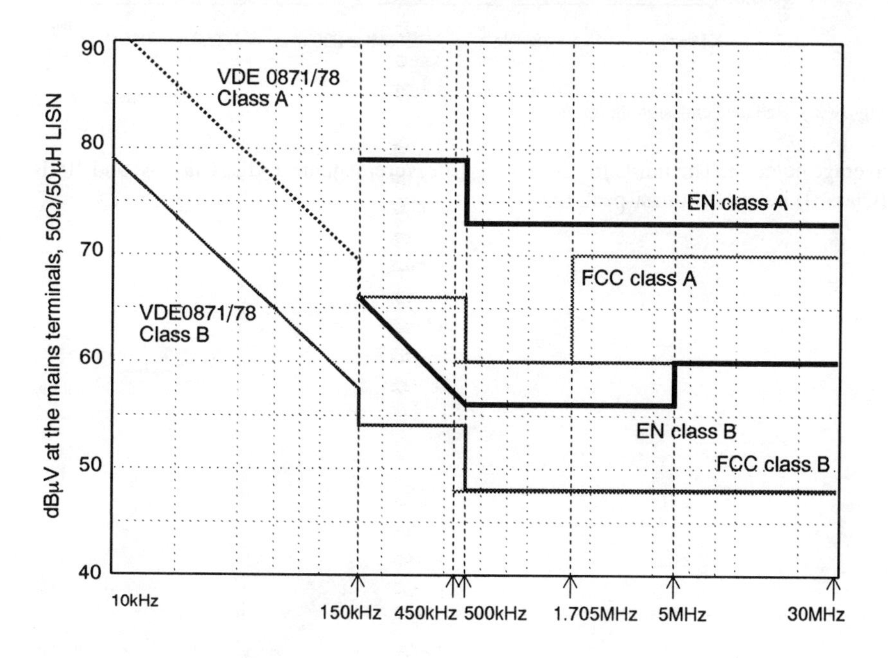

Figure 2.2 Conducted emission limits

For the purposes of these figures, EN class A refers to EN 55 011, EN 55 022 Class A and pr EN 50 081-2, and EN class B refers to EN 55 011, EN 55 022 Class B, EN 55 013, EN 55 014 (appliances, conducted only), and EN 50 081-1. All values are measured with the CISPR 16 quasi-peak detector, but note that EN 55 011, EN 55 013, EN 55 022 and EN 50 081 also require conducted emissions to be measured with an

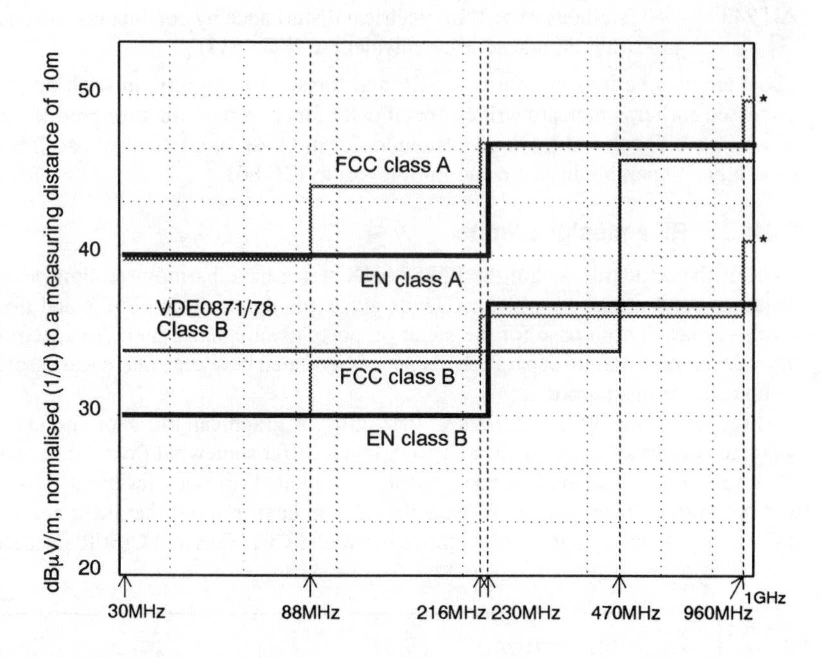

Figure 2.3 Radiated emission limits

average detector. The limits for the average measurement are 13dB (Class A) and 10dB (Class B) below the quasi-peak limits.

Chapter 3

EMC Measurements

One of the aspects of electromagnetic compatibility that is most difficult to grasp is the raft of techniques that are involved in making measurements. EMC phenomena extend in frequency to well beyond 1GHz and this makes conventional and well-known techniques, established for low frequency and digital work, quite irrelevant. Development and test engineers must appreciate the basics of high frequency measurements in order to perform the EMC testing that will be demanded of them. This chapter will serve as an introduction to the equipment, the test methods and some of the causes of error and uncertainty that attend EMC testing.

3.1 RF emissions testing

For ease of measurement and analysis, radiated emissions are assumed to predominate above 30MHz and conducted emissions are assumed to predominate below 30MHz. There is of course no magic changeover at 30MHz. But typical cable lengths tend to resonate above 30MHz, leading to anomalous conducted measurements, while measurements of radiated fields below 30MHz will of necessity be made in the near field if closer to the source than $\lambda/2\pi$ (see section 4.1.3.2), which gives results that do not necessarily correlate with real situations. In practice, investigations of interference problems have found that reduction of the noise voltages developed at the mains terminals has been successful in alleviating radio interference in the long, medium and short wave bands [48]. At higher frequencies, mains wiring becomes less efficient as a propagation medium and the dominant propagation mode becomes radiation from the equipment or wiring in its immediate vicinity.

Emissions testing requires that the equipment under test (EUT) is set up within a controlled electromagnetic environment under its normal operating conditions. If the object is to test the EUT alone, rather than as part of a system, its associated equipment (if any) must be separately screened from the measurement. Any ambient signals should be well below the levels to which the equipment will be tested.

The operating configuration is normally specified within emissions standards to be that which maximizes emissions. This is not always easy to predict and you may have to perform some preliminary tests while varying the configuration. Also, one configuration may generate high emissions in one part of the spectrum and another configuration may generate a different set of high emissions. It is the manufacturer's responsibility to specify the operating conditions that will be tested.

3.1.1 Measurement instrumentation

3.1.1.1 Measuring receiver

Conformance test measurements are normally taken with a measuring receiver. These are in most cases too expensive for ordinary companies' development labs and they are usually only found in EMC test houses. They are optimised for the purpose of taking EMC measurements. Typical costs for a complete receiver system to cover the range 10kHz to 1GHz are of the order of £40–60,000.

Early measuring receivers were manually tuned and the operator had to take readings from the meter display at each frequency that was near to the limit line. This was a lengthy procedure and prone to error. The current generation of receivers are fully automated and can be software controlled via an IEEE-488 standard bus; this allows a PC-resident program to take measurements with the correct parameters over the full frequency range of the test, in the minimum time consistent with gap-free coverage. Results are stored in the PC's memory and can be processed or plotted at will.

The distinguishing features of a measuring receiver compared to a spectrum analyser are:

- lack of a wide spectrum display for instantaneous diagnostics – the receiver output is provided at a spot frequency;
- very much better sensitivity, allowing signals to be discriminated from the noise at levels much lower than the emission limits;
- robustness of the input circuits, and resistance to overloading;
- intended specifically for measuring to CISPR standards, with bandwidths and detectors tailored for this purpose;
- frequency and amplitude accuracy better than the cheaper spectrum analysers;
- normally two units are required, one covering up to 30MHz and the other covering 30 – 1000MHz.

3.1.1.2 Spectrum analyser

A fairly basic spectrum analyser is considerably cheaper than a measuring receiver (typically £10 – 15,000) and is widely used for "quick-look" testing and diagnostics. The instantaneous spectrum display is extremely valuable for confirming the frequencies and nature of offending emissions, as is the ability to narrow-in on a small part of the spectrum. When combined with a tracking generator, a spectrum analyser is useful for checking the HF response of circuit networks.

Basic spectrum analysers are not an alternative to a measuring receiver because of their limited sensitivity and dynamic range, and susceptibility to overload. Figure 3.1(a) shows the block diagram of a typical spectrum analyser. The input signal is fed straight into a mixer which covers the entire frequency range of the analyser with no advance selectivity or preamplification. The consequences of this are threefold: firstly, the noise figure is not very good, so that when the attenuation due to the transducer and cable is taken into account, the sensitivity is hardly enough to discriminate signals from noise at the lower emission limits. Secondly, the mixer diode is a very fragile component and is easily damaged by momentary transient signals or continuous overloads at the input. If you take no precautions to protect the input, you will find your repair bills escalating quickly. Thirdly, the energy contained in broadband signals can overload the mixer and drive it into non-linearity even though the energy within the detector bandwidth is within the instrument's apparent dynamic range.

Preselector

You can find instruments which offer a performance equivalent to that of a measuring

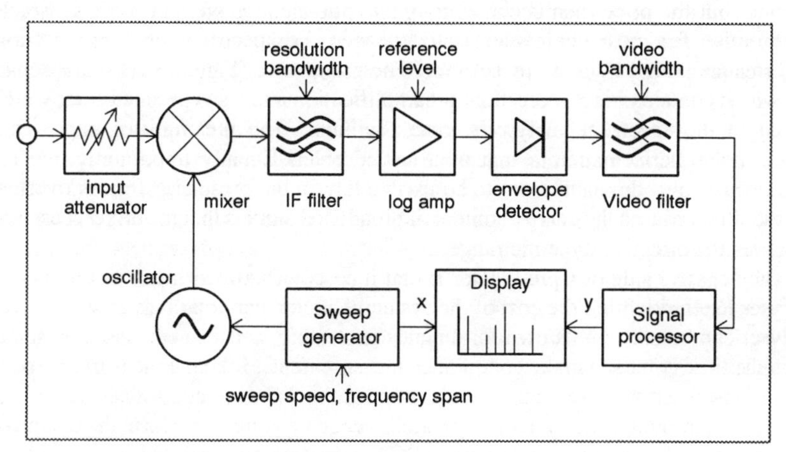

a) spectrum analyser

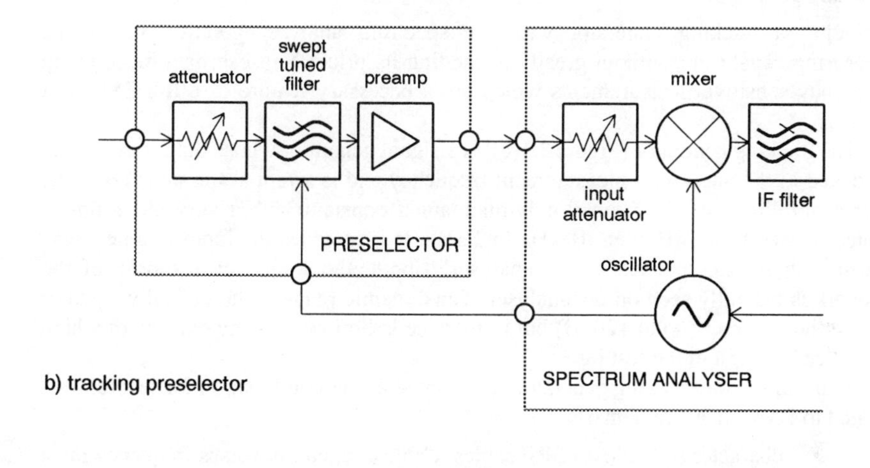

b) tracking preselector

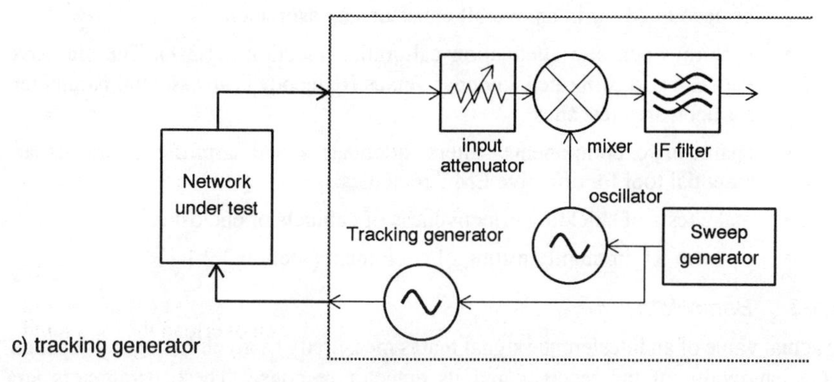

c) tracking generator

Figure 3.1 Block diagram of spectrum analyser

receiver, but the price then becomes roughly equivalent as well. A more satisfactory compromise for most companies is to enhance the spectrum analyser's front-end performance with a tracking preselector. The preselector (Figure 3.1(b)) is a separate unit which contains input protection, preamplification and a swept tuned filter which is locked to the spectrum analyser's local oscillator. The preamplifier improves the system noise performance to that of a test receiver. Equally importantly, the input protection allows the instrument to be used safely in the presence of gross overloads, and the filter reduces the energy content of broadband signals that the mixer sees, which improves the effective dynamic range.

The negative side of a preselector is that it costs virtually as much as the spectrum analyser itself, doubling the cost of the system. But you can treat it as an upgrade. The analyser can be used on its own for diagnostics testing, and you can add a preselector when the time comes to make compliance measurements. Like the measuring receiver, modern spectrum analysers and preselectors can be software controlled via an IEEE-488 bus, and provided the system hardware is adequate you can perform the compliance testing in exactly the same way (using a PC for the data processing).

Tracking generator

Including a tracking generator with the spectrum analyser greatly expands its measuring capabillity without greatly expanding its price. You can then make many frequency-sensitive measurements which are a necessary feature of a full EMC test facility.

The tracking generator (Figure 3.1(c)) is a signal generator whose output frequency is locked to the analyser's measurement frequency and is swept at the same rate. The output amplitude of the generator is maintained constant within very close limits, typically less than 1dB over 100kHz to 1GHz. If it provides the input to a network whose output is connected to the analyser's input, the frequency response of the network is instantly seen on the analyser. The dynamic range is theoretically equal to that of the analyser (up to 120dB) but in practice is limited by stray reactances which cause feedthrough in the test jig.

You can use the tracking generator/spectrum analyser combination for several tests related to EMC measurements:

- characterise the loss of RF cables. Cable attenuation versus frequency must be accounted for in an overall emissions measurement
- perform open site attenuation calibration (section 3.1.3.1). The site loss between two calibrated antennas versus frequency is an essential parameter for open area test sites
- characterise components, filters, attenuators and amplifiers. This is an essential tool for effective EMC remedies
- make tests of shielding effectiveness of cabinets or enclosures
- perform RF immunity testing of equipment (section 3.2.1.1)

3.1.1.3 Bandwidth

The actual value of an interference signal that is measured at a given frequency depends on the bandwidth of the receiver and its detector response. These parameters are rigorously defined in a separate standard that is referenced by all the commercial emissions standards that are based on the work of CISPR, notably EN55011, 55013, 55014 and 55022. This standard is CISPR16 [102], aligned in the UK with BS727.

CISPR16 splits the measurement range of 9kHz to 1000MHz into four bands, and defines a measurement bandwidth for quasi-peak detection which is constant over each of these bands (Table 3.1). Sources of emissions can be classified into narrowband, usually due to oscillator harmonics, and broadband, due to discontinuous switching operations and digital data transfer. The actual distinction between narrowband and broadband is based on the bandwidth occupied by the signal compared with the bandwidth of the measuring instrument. A broadband signal is one whose occupied bandwidth exceeds that of the measuring instrument. Thus a signal with a bandwidth of 30kHz at 20MHz (CISPR band B) would be classed as broadband, while the same signal at 40MHz (band C) would be classed as narrowband.

	Frequency band			
	A 10–150kHz	B 0.15–30MHz	C 30–300MHz	D 300–1000MHz
Quasi-peak detector				
6dB bandwidth, kHz	0.2	9	120	
Charge time constant, ms	45	1	1	
Discharge time constant, ms	500	160	550	
Pre-detector overload factor, dB	24	30	43.5	

Table 3.1 The CISPR 16 quasi peak detector and bandwidths

Noise level versus bandwidth

The indicated level of a broadband signal changes with the measuring bandwidth. As the measuring bandwidth increases, more of the signal is included within it and hence the indicated level rises. The indicated level of a narrowband signal is not affected by measuring bandwidth. Noise, of course, is inherently broadband and therefore there is a direct correlation between the "noise floor" of a receiver or spectrum analyser and its measuring bandwidth: minimum noise (maximum sensitivity) is obtained with the narrowest bandwidth. The relationship between noise and bandwidth is given by equation (3.1):

$$\text{Noise level change (dB)} = 10\log_{10}(BW_1/BW_2) \qquad (3.1)$$

3.1.1.4 Detector function

There are three kinds of detector in common use in RF emissions measurements: peak, quasi peak and average. The characteristics are defined in CISPR16 (see Table 3.1) and are different for the different frequency bands.

By no means all interference emissions are continuous at a fixed level. A carrier signal may be amplitude modulated, and both a carrier and a broadband emission may be pulsed. The measured level which is indicated for different types of modulation will depend on the type of detector in use. Figure 3.2 shows the indicated levels for the three detectors with various modulation shapes.

Peak

The peak detector responds near-instantaneously to the peak value of the signal and discharges fairly rapidly. If the receiver dwells on a single frequency the peak detector

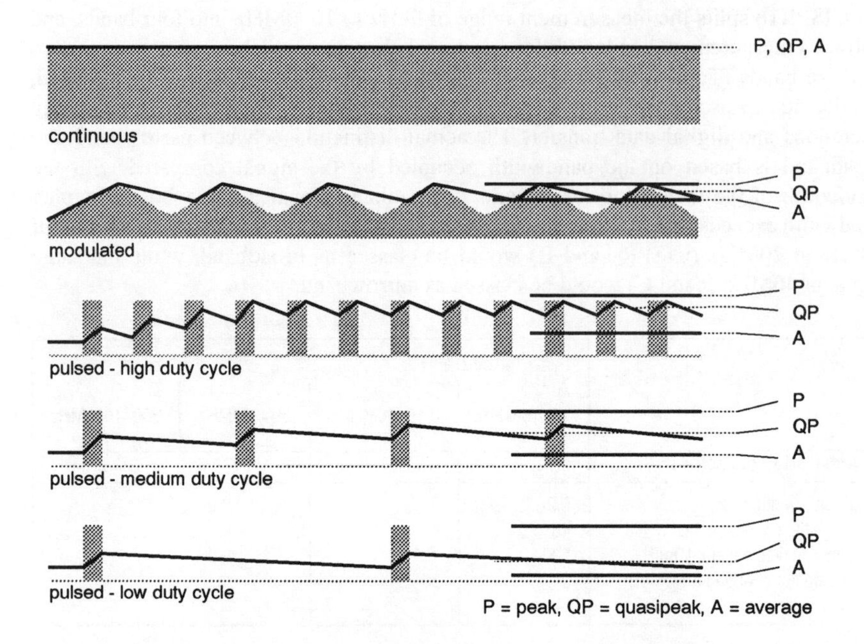

Figure 3.2 Indicated level versus modulation waveform for different detectors

output will follow the "envelope" of the signal, hence it is sometimes called an envelope detector. Military specifications make considerable use of the peak detector, but CISPR emissions standards do not require it at all. However its fast response makes it very suitable for diagnostic or "quick-look" tests, and it can be used to speed up a proper compliance measurement as is outlined in section 3.1.4.3.

Average

The average detector, as its name implies, measures the average value of the signal. For a continuous signal this will be the same as its peak value, but a pulsed or modulated signal will have an average level lower than the peak. EN55022 [91] and its derivative standards call for an average detector measurement on conducted emissions, with limits which are 10-13dB lower than the quasi-peak limits. The effect of this is to penalise continuous emissions with respect to pulsed interference, which registers a lower level on an average detector [49]. A simple way to make an average measurement on a spectrum analyser is to reduce the post-detector "video" bandwidth to well below the lowest expected modulation or pulse frequency [53].

Quasi-peak

The quasi-peak detector is a peak detector with weighted charge and discharge times (Table 3.1) which correct for the subjective human response to pulse-type interference. Interference at low pulse repetition frequencies (PRF) is subjectively less annoying on radio reception than that at high PRFs. Therefore, the quasi-peak response de-emphasizes the peak response at low PRFs.

3.1.1.5 Overload factor

A pulsed signal with a low duty cycle, measured with a quasi-peak or average detector, should show a level that is less than its peak level by a factor which depends on its duty cycle and the relative time constants of the quasi-peak detector and PRF. To obtain an accurate measurement the signal that is presented to the detector must be undistorted at very much higher levels than the output of the detector. The lower the PRF, the higher will be the peak value of the signal for a given output level (Figure 3.3). Conventionally, the input attenuator is set to optimise the signal levels through the receiver, but the required pulse response means that the RF and IF stages of the receiver must be prepared to be overloaded by up to 43.5dB (for CISPR bands C and D) and remain linear. This is an extremely challenging design requirement and partially accounts for the high cost of proper measuring receivers, and the unsuitability of spectrum analysers for pulse measurements.

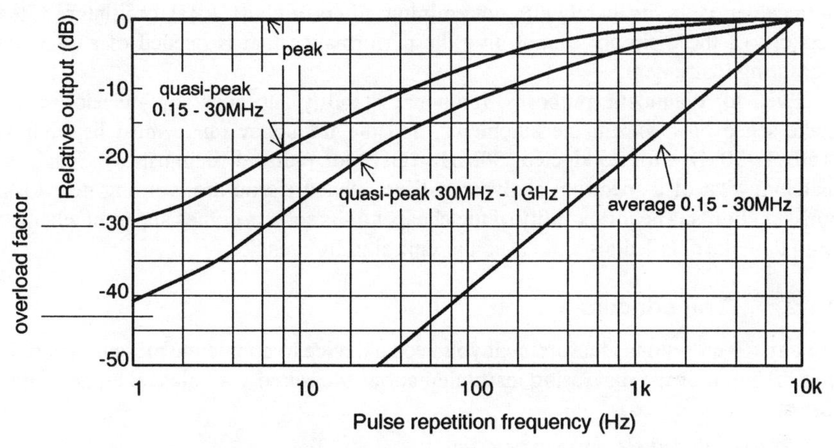

Figure 3.3 Relative output versus PRF for CISPR 16 detectors

The same problem means that the acceptable range of PRFs that can be measured by an average detector is limited. The overload factor of receivers up to 30MHz is only required to be 30dB, and this degree of overload would be reached on an average detector with a pulsed signal having a PRF of less than 300Hz. For this reason average detectors are only intended for measurement of continuous signals to allow for modulation or the presence of broadband noise, and are not generally used to measure impulsive interference.

3.1.1.6 Measurement time

Both the quasi-peak and the average detector require a relatively long time for their output to settle on each measurement frequency. This time depends on the time constants of each detector and is measured in hundreds of milliseconds. When a range of frequencies is being measured, the conventional method is to step the receiver at a step size of around half its measurement bandwidth, in order to cover the range fully without gaps. For a complete measurement scan of the whole frequency range, as is required for a compliance test, the time taken is given by

$$T = (\text{frequency span}/0.5 \cdot \text{bandwidth}) \cdot \text{dwell time per spot frequency} \qquad (3.2)$$

If the dwell time is restricted to three time constants, the time taken to do a complete quasi-peak sweep from 150kHz to 30MHz turns out to be 53 minutes. For an average measurement with an even longer dwell time the scan time is correspondingly longer. If the signal being measured is modulated or pulsed, the dwell time must be increased to ensure that the peaks are captured. This has repercussions on the test method, as is discussed later in section 3.1.4. It places correspondingly severe restrictions on the sweep rate when you are using a spectrum analyser [24].

3.1.1.7 Other measuring instruments

Instruments are now appearing on the market which fulfil some of the functions of a spectrum analyser at a much lower price. Prominent among these are units which convert an oscilloscope into a spectrum display. Such devices are useful for diagnostic purposes provided that you recognise their limitations – typically frequency range, stability, bandwidth and/or sensitivity. The major part of the cost of a spectrum analyser or receiver is in its bandwidth-determining filters and its local oscillator. Cheap versions of these simply cannot give the performance that is needed of an accurate measuring instrument.

Even for diagnostic purposes, frequency stability and accuracy are necessary to make sense of spectrum measurements, and the frequency range must be adequate (150kHz-30MHz for conducted, 30MHz-1GHz for radiated diagnostics). Sensitivity matching that of a spectrum analyser will be needed if you are working near to the emission limits. The inflexibility of the cheaper units soon becomes apparent when you want to make detailed tests of particular emission frequencies.

3.1.2 Transducers

For any RF emissions measurement you need a device to couple the measured variable into the input of the measuring instrumentation. Measured variables take one of three forms:

- radiated electromagnetic field
- conducted cable voltage
- conducted cable current

and the transducers for each of these forms are discussed below.

3.1.2.1 Antennas

The basics of electromagnetic fields are outlined in section 4.1.3.1. Radiated field measurements can be made of either electric (E) or magnetic (H) field components. In the far field the two are equivalent, and related by the impedance of free space:

$$E/H = Z_0 = 120\pi = 377\Omega \qquad (3.3)$$

but in the near field they are unrelated. In either case, an antenna is needed to couple the field to the measuring receiver. Electric field strength limits are specified in terms of volts (or microvolts) per metre at a given distance from the EUT, whilst measuring receivers are calibrated in volts (or microvolts) at the 50Ω input. The antenna must therefore be calibrated in terms of volts output into 50Ω for a given field strength at each frequency; this calibration is known as the antenna factor.

Most standards allow the use of broadband antennas, which obviate the need for re-tuning at each frequency. The two most common broadband devices are the biconical, for the range 30 – 300MHz, and the log periodic, for the range 300 – 1000MHz. Some antennas have different frequency ranges, but it is always possible to combine a biconical and a log periodic to cover the range 30-1000MHz. A tuned dipole is also specified as an alternative and has the advantage that its performance can be accurately predicted, but because it can only be applied at spot frequencies it is not used for everyday measurement but is reserved for calibration of broadband antennas, site surveys, site attenuation measurements and other more specialized purposes.

Antenna parameters

Those who use antennas for radio communication purposes are familiar with the specifications of gain and directional response, but these are of only marginal importance for EMC measurements. The antenna is always oriented for maximum response. Antenna factor is the most important parameter, and each calibrated broadband antenna is supplied with a table of its antenna factor versus frequency. Typical antenna factors for a biconical and a log periodic are shown in Figure 3.4. To convert the measured voltage at the instrument terminals into the actual field strength at the antenna you have to add the antenna factor and cable attenuation (Figure 3.5). Cable attenuation is also a function of frequency.

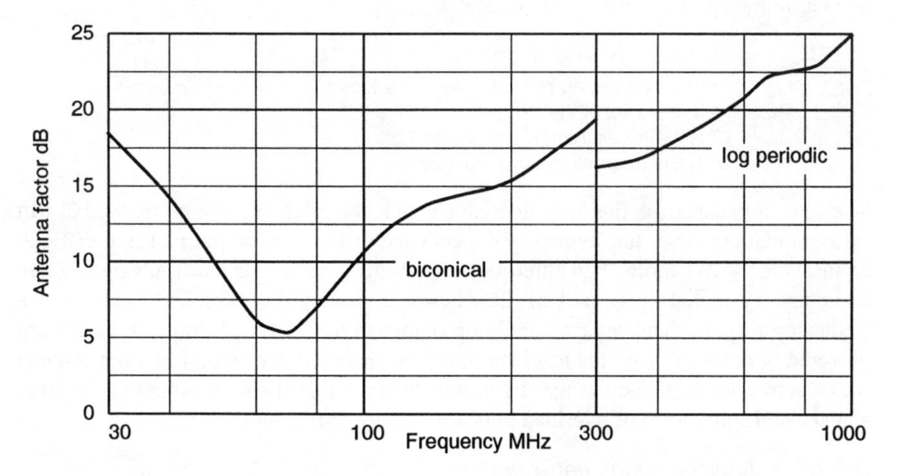

Figure 3.4 Typical antenna factors

In the far field the electric and magnetic fields are orthogonal (Appendix C, section C.3). With respect to the physical environment each field may be vertically or horizontally polarized, or in any direction in between. The actual polarization depends on the nature of the emitter and on the effect of reflections from other objects. An antenna will show a maximum response when its plane of polarization aligns with that of the incident field, and will show a minimum when the planes are at right angles. The plane of polarization of both biconical and log periodic is in the plane of the elements.

The loop antenna

The majority of radiated emissions are measured in the range 30 to 1000MHz. A few standards call for radiated measurements below 30MHz. In these cases the magnetic field strength is measured, using a loop antenna. Measurements of the magnetic field give better repeatability in the near field than of the electric field, which is easily perturbed by nearby objects. The loop is merely a coil of wire which produces a voltage

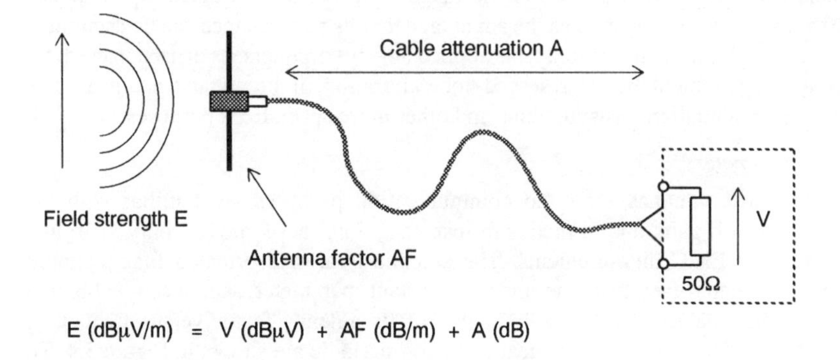

Figure 3.5 Converting field strength to measured voltage

at its terminals proportional to frequency:

$$E = 4\pi \cdot 10^{-7} \cdot N \cdot A \cdot 2\pi F \cdot H \quad (3.4)$$

where N is the number of turns in the loop
A is the area of the loop, m^2
F is the measurement frequency, Hz
H is the magnetic field, Amps/metre

The low impedance of the loop does not match the 50Ω impedance of typical test instrumentation. Also, the frequency dependence of the loop output makes it difficult to measure across more than three decades of frequency. These disadvantages are overcome by including as part of the antenna a preamplifier which corrects for the frequency response and matches the loop output to 50Ω. The preamp can be battery powered or powered from the test instrument. Such an "active" loop has a flat antenna factor across its frequency range. Its disadvantage is that it can be saturated by large signals, and some form of overload indication is needed to warn of this.

3.1.2.2 Artificial mains network

To make conducted voltage emissions tests on the mains port, you need an artificial mains network or Line Impedance Stabilising Network (LISN) to provide a defined impedance at RF across the measuring point, to couple the measuring point to the test instrumentation and to isolate the test circuit from unwanted interference signals on the supply mains. The most widespread type of LISN is defined in CISPR 16 and presents an impedance equivalent to 50Ω in parallel with 50μH across each line to earth (Figure 3.6). Others are also defined, but the 50Ω/50μH version has become the norm.

Note that its impedance is not defined above 30MHz, partly because commercial conducted measurements are not required above this frequency (aerospace and automotive standards do call for conducted tests above 30MHz) but also because component parasitic reactances make a predictable design difficult to achieve. CISPR 16 includes a suggested circuit (Figure 3.7) for each line of the LISN, but it only actually defines the impedance characteristic. The main impedance determining components are the measuring instrumentation input impedance, the 50μH inductor and the 5Ω resistor. The remaining components serve to decouple the incoming supply.

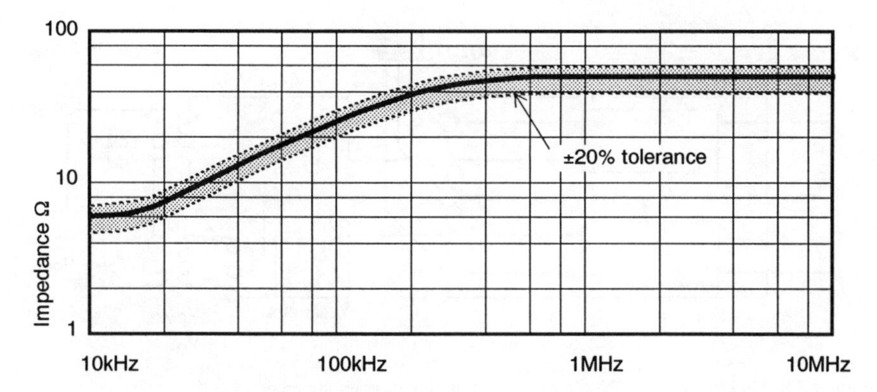

Figure 3.6 LISN impedance versus frequency

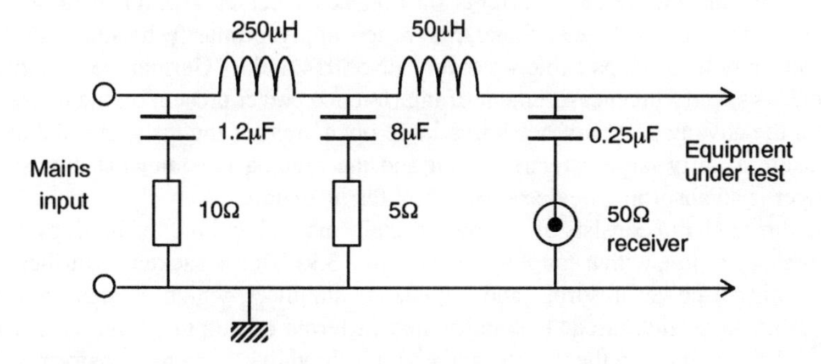

Figure 3.7 LISN circuit

Earth current

A large capacitance (in total approaching 10μF) is specified between line and earth, which when exposed to the 240V line voltage results in around 0.75A in the safety earth. This level of current is lethal, and the unit must therefore be solidly connected to earth for safety reasons. If it is not, the LISN case, the measurement signal lead and the equipment under test (EUT) can all become live. As a precaution, you are advised to bolt your LISN to a permanent ground plane and not allow it to be carried around the lab! A secondary consequence of this high earth current is that LISNs cannot be used on mains circuits that are protected by earth leakage or residual current circuit breakers.

Diagnostics with the LISN

As it stands, the LISN does not distinguish between differential mode (line to line) and common mode (line to earth) emissions (see section 4.2.1.3 and section 4.2.2); it merely connects the measuring instrument between either live or neutral and earth. A modification to the LISN circuit (Figure 3.8) allows you to detect either the sum or the difference of the live and neutral voltages, which correspond to the common mode and differential mode voltages respectively [69]. This is not required for compliance measurements but is very useful when making diagnostic tests on mains filters.

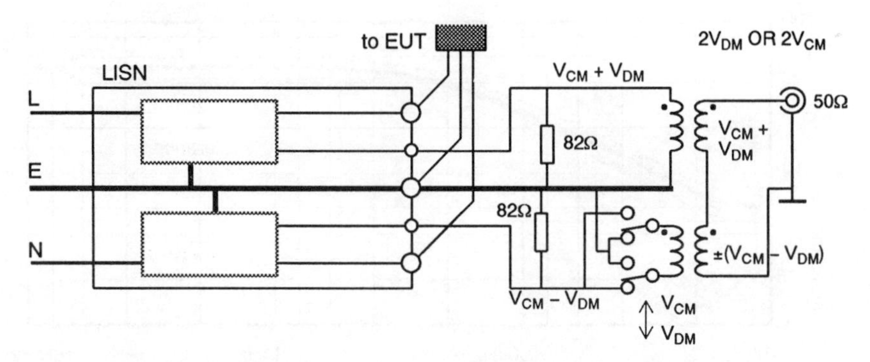

Figure 3.8 Modifying the LISN to measure differential or common mode

3.1.2.3 Absorbing clamp and current probe

As well as measuring the emissions above 30MHz directly as a radiated field you can also measure the interference currents on connected cables and relate these to the accepted field strength levels. Standards which apply primarily to small apparatus connected only by a mains cable – notably EN55014 and its German near-equivalent VDE0875 – specify the measurement of interference power present on the mains lead. This has the advantage of not needing a large open area for the tests, but it should be done inside a fairly large screened room and the method is somewhat clumsy. The transducer is an absorbing device known as a ferrite clamp.

The ferrite clamp consists of a current transformer using three ferrite rings, split to allow cable insertion, with a coupling loop (Figure 3.9). This is backed by further ferrite rings forming a power absorber and impedance stabiliser, which clamps around the mains cable to be measured. It is calibrated in terms of output power versus input power. The purpose of the ferrite absorbers behind the current transformer is to attenuate reflections and extraneous signals that would otherwise appear at the current transformer. The lead from the current transformer to the measuring instrument is also sheathed with ferrite rings to attenuate screen currents on this cable. Because the output is proportional to current flowing in common mode on the measured cable, it can be used as a direct measure of noise power, and the clamp can be calibrated as a two-port network in terms of output power versus input power.

CISPR 16 specifies the construction, calibration and use of the ferrite clamp. As well as its use in certain compliance tests, it also lends itself to diagnostics as it can be used for repeatable comparative measurements on a single cable to check the effect of circuit changes.

Current probe

The same applies to the current probe, which is essentially the same as the absorbing clamp except that it doesn't have the absorbers. It is simply a clamp-on, calibrated wideband current transformer. Military specifications call for its use on individual cable looms, and a draft amendment to EN 55022 proposing test methods for signal ports also requires a current probe. CISPR 16 includes a specification for the current probe. Because the current probe does not have an associated absorber, the RF common mode termination impedance of the line under test must be defined by an impedance stabilising network, which must be transparent to the signals being carried on the line.

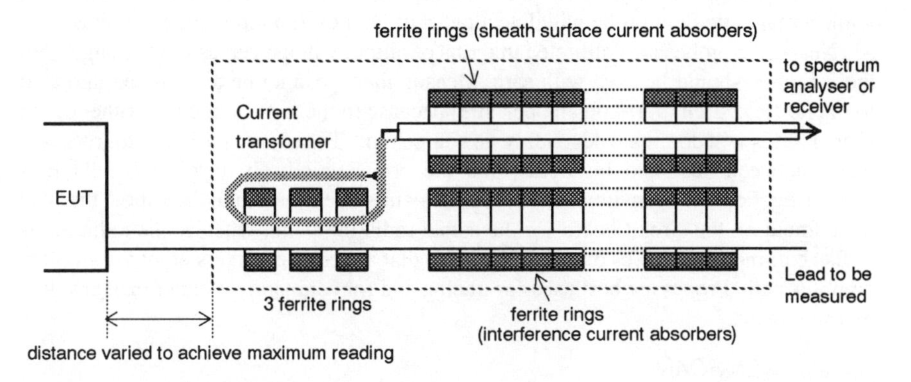

Figure 3.9 The absorbing ferrite clamp

Both the ferrite clamp and the current probe have the great advantage that no direct connection is needed to the cable under test, and disturbance to the circuit is minimal since the probe effect is no more than a slight increase in common mode impedance.

3.1.2.4 Near field probes

Very often you will need to physically locate the source of emissions from a product. A set of near-field (or "sniffer") probes is used for this purpose. These are so-called because they detect field strength in the near field, and therefore two types of probe are needed, one for the electric field (rod construction) and the other for the magnetic field (loop construction). It is simple enough to construct adequate probes yourself using coax cable (Figure 3.10), or you can buy a calibrated set. A probe can be connected to a spectrum analyser for a frequency domain display, or to an oscilloscope for a time domain display.

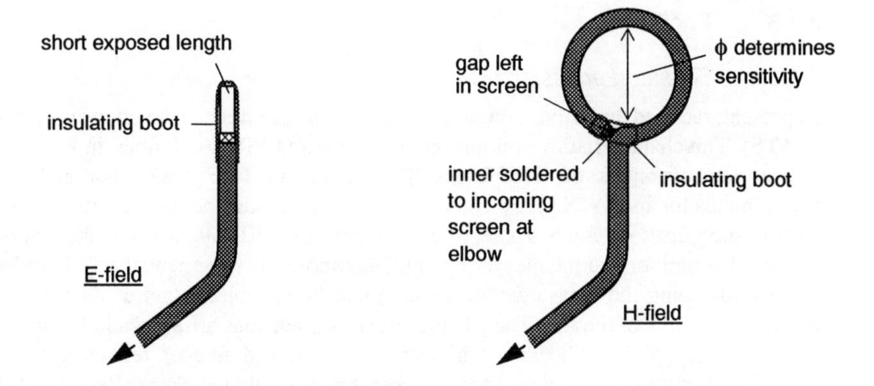

Figure 3.10 Do-it-yourself near field probes

Probe design is a trade-off between sensitivity and spatial accuracy. The smaller the probe, the more accurately it can locate signals but the less sensitive it will be. You can increase sensitivity with a preamplifier if you are working with low-power circuits. A good magnetic field probe is insensitive to electric fields and vice versa; this means that an electric field probe will detect nodes of high dv/dt but will not detect current paths, while a magnetic field probe will detect paths of high di/dt but not voltage points.

Near-field probes are calibrated in terms of output voltage versus field strength, but these figures should be used with care. Measurements cannot be directly extrapolated to the far field strength (as on an open site) because in the near field one or other of the E or H fields will dominate depending on source type. The sum of radiating sources will differ between near and far fields, and the probe will itself distort the field it is measuring. Perhaps more importantly, you may mistake a particular "hot spot" that you have found on the circuit board for the actual radiating point, whereas the radiation is in fact coming from cables or other structures that are coupled to this point via an often complex path. Probes are best used for tracing and for comparative rather than absolute measurements.

3.1.2.5 EMSCAN

One further device which should be mentioned here is the EMSCAN™†. This was developed and patented at Bell Northern Research in Canada and is now beginning to be marketed in Europe and Japan. Its price at present puts it out of reach of most companies but this will probably fall in the future. In principle it is essentially a planar array of tiny near field current probes arranged in a grid form on a multilayer pcb [34]. The output of each current probe can be switched under software control to a frequency selective measuring instrument, whose output in turn provides a graphical display on the controlling workstation.

The device is used to provide a near-instantaneous two dimensional picture of the RF emissions directly from a printed circuit card placed over the scanning unit. It can provide either a frequency versus amplitude plot of the near field at a given location on the board, or an x-y co-ordinate map of the current distribution at a given frequency. For the designer it can instantly show the effect of remedial measures on the pcb being investigated, while for production quality assurance it can be used to evaluate batch samples which can be compared against a known good standard.

3.1.3 Facilities

3.1.3.1 Radiated emissions

At present, radiated emissions compliance testing must be done on an open area test site (OATS). The characteristics of a minimum standard OATS are defined in EN 55 022 and in later versions of CISPR 16. The American FCC have also laid down requirements for an OATS [127] which differ in detail though not in substance from the CISPR specification. Such a site offers a controlled RF attenuation characteristic between the emitter and the measuring antenna (known as site attenuation). In order to avoid influencing the measurement there should be no objects that could reflect RF within the vicinity of the site. The CISPR test site dimensions are shown in Figure 3.11.

The ellipse defines the area which must be flat and free of reflecting objects, including overhead wires. In practice, for good repeatability between different test sites a substantially larger surrounding area free from reflecting objects is advisable. This means that the room containing the control and test instrumentation needs to be some distance away from the site. An alternative is to put this room directly below the ground plane, either by excavating an underground chamber or by using the flat roof of an existing building as the test site.

† EMSCAN is a trademark of Northern Telecom, Canada

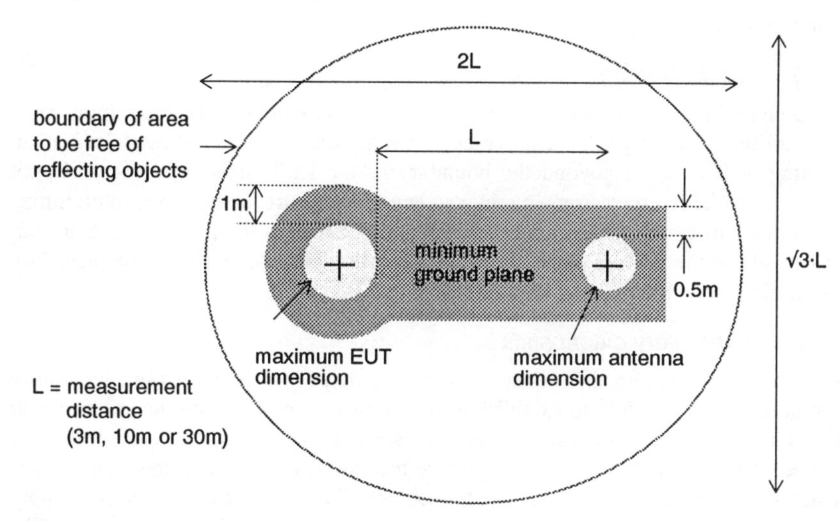

Figure 3.11 The CISPR OATS

Ground plane

Because it is impossible to avoid ground reflections, these are regularized by the use of a ground plane. The minimum ground plane dimensions are also shown in Figure 3.11. Again, an extension beyond these dimensions will improve repeatability, as will close attention to the construction of the ground plane. It should preferably be solid metal sheets welded together, but this may be impractical. Joints or mesh should not have voids or gaps that are greater than 0.1λ at the highest frequency (3cm). Ordinary wire mesh is unsuitable unless each individual overlap of the wires is bonded.

Measuring distance

The measurement distance between EUT and receiving antenna determines the overall dimensions of the site and hence its expense. There are three commonly specified distances, 3m, 10m and 30m. There is a proposal to standardize compliance measurements to 10m. Confidence checks can comfortably be carried out on a 3m range, on the assumption that levels measured at 10m will be 10dB lower (field strength is proportional to 1/d). This assumption is not entirely valid at the lower end of the frequency range, where 3m separation is approaching the near field.

Radiated measurements in a screened chamber

Although you can make radiated emissions measurements in a large anechoic screened chamber (see section 3.2.2.1) at higher frequencies, use of these is not expressly permitted in current standards. Alternative sites to those specified above are permitted provided that errors due to their use do not invalidate the results. A draft standard has been circulated by CENELEC [97] defining requirements for screened anechoic enclosures for emission measurement, and in due course it is expected that test methods using sites defined within this standard will be allowed within product standards. The major requirement is to establish that the site attenuation is not significantly affected by wall reflections and the draft standard proposes a method for doing this. Wall reflections from a non-anechoic (unlined) screened chamber will severely distort the site attenuation, and accurate measurements of radiated emissions in such a chamber are impossible.

3.1.3.2 Conducted emissions

By contrast with radiated emissions, conducted measurements need the minimum of extra facilities. The only vital requirement is for a ground plane of at least 2m by 2m, extending at least 0.5m beyond the boundary of the EUT. It is convenient but not essential to make the measurements in a screened enclosure, since this will minimize the amplitude of extraneous ambient signals, and one wall or the floor of the room can then be used as the ground plane. Non-floor-standing equipment should be placed on an insulating table 40cm above the ground plane.

3.1.3.3 Laboratory diagnostics

Although you will not be able to make accurate radiated measurements in a laboratory environment, it is possible to establish a minimum set-up in one corner of the lab at which you can perform emissions diagnostics and carry out comparative tests. For example, if you have done a pre-compliance test at a test house and have discovered one particular frequency at 10dB above the required limit, back in the lab you can apply remedial measures and check each one to see if it gives you a 15dB improvement (5dB margin) without being concerned for the absolute accuracy. While this method is not absolutely foolproof, it is often the best that companies with limited resources and facilities can do.

The following checklist suggests a minimum set-up for doing this kind of in house diagnostic work.

- unrestricted floor area of at least 5m x 3m to allow a 3m test range with 1m beyond the antenna and EUT;
- this floor area to be permanently covered with a ground plane of aluminium or copper sheet or foil bonded together and to safety earth;
- no other electronic equipment which could generate extraneous emissions (especially computers) in the vicinity, the EUT's support equipment should be well removed from the test area;
- no mobile reflecting objects in the vicinity, or those which are mobile should have their positions carefully marked for repeatability;
- an insulating table or workbench at one end of the test range on which to put the EUT, with a LISN bonded to the ground plane beneath it;
- equipment consisting of a spectrum analyser, limiter, antenna set and insulating tripod;
- antenna polarization maintained at horizontal, since this reduces errors due to reflections and ground proximity.

Once this set-up is established it should not be altered between measurements on a given EUT. This will give you a reasonable chance of repeatable measurements even if their absolute accuracy cannot be determined.

3.1.4 Test methods

The major part of all the standards referred to in chapter 2 consists of recipes for carrying out the tests. Because the values obtained from measurements at RF are so dependent on layout and method, these have to be specified in some detail to generate a standard result. This section summarizes the issues involved, but to actually perform the tests you are recommended to consult the relevant standard carefully.

3.1.4.1 Layout

For conducted emissions, the principal requirement is placement of the EUT with respect to the ground plane and the LISN, and the disposition of the mains cable and earth connection(s). A complication is that the specified layout for CISPR tests and that for FCC tests differ in these details, so that tests to the two specifications are not exactly equivalent. Figure 3.12 shows the layout for conducted emissions testing.

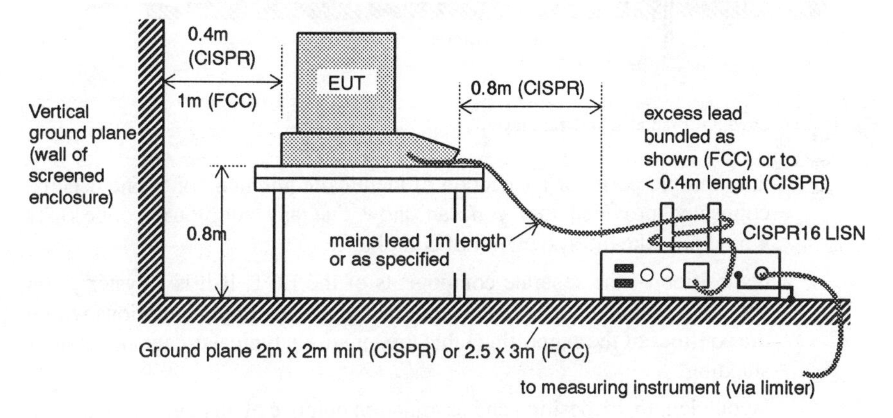

Figure 3.12 Layout for conducted emission tests

Radiated emissions require the EUT to be positioned so that its boundary is the specified distance from the measuring antenna. "Boundary" is defined as "an imaginary straight line periphery describing a simple geometric configuration" which encompasses the EUT. A non-floor-standing EUT should be 0.8m above the ground plane. The EUT will need to be rotated through 360° to find the direction of maximum emission, and this is usually achieved by standing it on a turntable. If it is too big for a turntable, then the antenna must be moved around the periphery while the EUT is fixed. Figure 3.13 shows the general layout for radiated tests.

3.1.4.2 Configuration

Once the date for an EMC test approaches, the question most frequently asked of test house engineers is "what system should I test?" The configuration of the EUT itself is not well specified in the current version of EN 55 022 (which is regarded as the root standard for emissions testing), although a proposed revision to EN 55 022 will rectify this. By contrast FCC document MP-4 [126] specifies both the layout and composition of the EUT in great detail, especially if the EUT is a personal computer or peripheral. Factors which will affect the emissions profile from the EUT, and which if not specified in the chosen standard should at least be noted in the test report, are

- number and selection of ports connected to ancillary equipment: you must decide on a "typical configuration". Where several different ports are provided each one should be connected to ancillary equipment. Where there

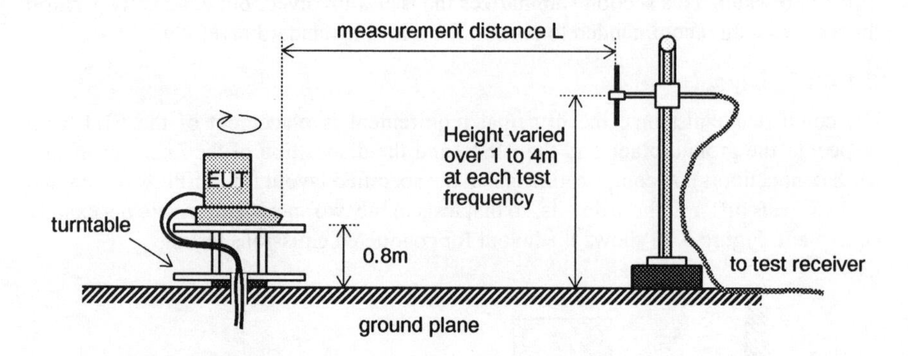

Figure 3.13 Layout for radiated emission tests

are multiple ports for connection of identical equipment, only one need be connected provided that you can show that any additional connections would not take the system out of compliance

- disposition of the separate components of the EUT, if it is a system; you should experiment to find the layout that gives maximum emissions within the confines of the supporting table top, or within typical useage if it is floor standing
- layout, length, disposition and termination practice of all connecting cables; excess cable lengths should be bundled (not looped) near the centre of the cable with the bundle 30–40cm long. Lengths and types of connectors should be representative of normal installation practice
- population of plug-in modules, where appropriate; as with ancillary equipment, one module of each type should be included to make up a minimum representative system. Where you are marketing a system (such as a data acquisition unit housed in a card frame) that can take many different modules but not all at once, you may have to define several minimum representative systems and test all of them
- software and hardware operating mode; FCC MP-4 specifies at least two modes of operation, i.e. equipment powered on and awaiting data transfer, and sending/receiving data in typical fashion. You should also define displayed video on VDUs and patterns being printed on a printer
- use of simulators for ancillary equipment is permissible (except for FCC certification of computing devices) provided that its effects on emissions can be isolated or identified. Any simulator must properly represent the actual RF electrical characteristics of the real interface
- EUT grounding method: should be that specified in your installation instructions. If the EUT is intended to be operated ungrounded, it must be tested as such. If it is grounded via the safety earth (green and yellow) wire in the mains lead, this should be connected to the measurement ground plane at the mains plug (for conducted measurements, this will be automatic through the LISN).

The catch-all requirement in all standards is that the layout, configuration and operating mode *shall be varied so as to maximize the emissions.* This means some exploratory testing once the significant emission frequencies have been found, varying all of the above parameters – and any others which might be relevant – to find the maximum point. For a complex EUT or one made up of several interconnected subsystems this operation is time consuming. Even so, you must be prepared to justify the use of whatever final configuration you choose in the test report.

FCC MP-4: personal computers

The FCC requirements for testing personal computers and their peripherals are specified in some depth. The minimum test configuration for any PC or peripheral must include the PC, a keyboard, an external monitor, an external peripheral for a serial port and an external peripheral for a parallel port. If the PC is not equipped with any of these interface ports, the ports must be added to it for the puposes of testing. If it is equipped with more than the minimum interface requirements, peripherals must be added to all the interface ports except where these would require identical peripherals, and provided that the addition of identical peripherals would not affect the test results. If the PC has an auxiliary power outlet for its monitor, then measurements must be made with the monitor powered from both this outlet and then from a separate outlet. The support equipment for the EUT must not be modified to allow the EUT to achieve compliance – this can make life difficult for firms testing outside the US, who up till now have had to import FCC-certified products such as printers or VDUs specifically for the emissions tests so that the EUT's emissions are not swamped by those from its support equipment.

The software exercising routine for the PC is also specified. The software generates a complete line of a continuously repeating "H" pattern, and sends it independently to the video port, the parallel port, the serial port, and writes to and reads from the disk drives. Monitors should display the "H" pattern and printers should print it.

3.1.4.3 Test procedure

The procedure which is followed for an actual compliance test, once you have found the configuration which maximizes emissions, is straightforward if somewhat lengthy. Conducted emissions require a continuous sweep from 150kHz to 30MHz at a fixed bandwidth of 9kHz, once with a quasi-peak detector and once with an average detector (from 450kHz with a quasi peak detector only for FCC). If the average limits are met with the quasi-peak detector there is no need to perform the average sweep. Radiated emissions require only a quasi-peak sweep from 30MHz to 1GHz with 120kHz bandwidth, but there are complications: firstly, two antennas are needed and so it is customary to cover the range in two sweeps, one with the biconical and the other with the log periodic antenna.

Secondly, measurements must be made with the receiving antenna in both horizontal and vertical polarization. Thus in fact four sweeps are made, one in each polarization over each frequency range.

Maximizing emissions

But most importantly, for each significant emission frequency, i.e. where the measured level is within say 10dB of the limit, the EUT must be rotated to find the maximum emission direction *and* the receiving antenna must be scanned in height from 1 to 4m to find the maximum level. If there are many emission frequencies this can take a very long time. With a test receiver, automatic turntable and antenna mast under computer control, software can be written to perform the whole operation. This removes one source of operator error and reduces the test time, but not substantially.

A further difficulty arises if the operating cycle of the EUT is intermittent: say its maximum emissions only occur for a few seconds and it then waits for a period before it can operate again. Since the quasi peak or average measurement is inherently slow, with a dwell time at each frequency of hundreds of milliseconds, interrupting the sweep or the azimuth or height scan to synchronize with the EUT's operating cycle is necessary and this stretches the test time further. If it is possible to speed up the operating cycle to make it continuous, as for instance by running special test software, this is well worthwhile in terms of the potential reduction in test time.

Fast pre-scan

A partial way around the difficulties of excessive test time is to make use of the characteristics of the peak detector (see section 3.1.1.4 and 3.1.1.6). Because it responds instantaneously to signals within its bandwidth the dwell time on each frequency can be short, just a few milliseconds at most, and so using it will enormously speed up the sweep rate for a whole frequency scan. Its disadvantage is that it will overestimate the levels of pulsed or modulated signals (see Figure 3.2). This is a positive asset if it is used on a qualifying pre-scan in conjunction with computer data logging. The pre-scan with a peak detector will only take a few seconds and all frequencies at which the level exceeds some pre-set value lower than the limit can be recorded in a data file. These frequencies can then be measured individually, with a quasi peak and/or average detector, and subjecting each one to a height and azimuth scan. Provided there are not too many of these spot frequencies the overall test time will be significantly reduced, as there is no need to use the slow detectors across the whole frequency range. You must be careful, though, if the EUT emissions include pulsed narrowband signals with a relatively low repetition rate – some digital data emissions have this characteristic – that the dwell time is not set so fast that the peak detector will miss some emissions as it scans over them. The dwell time should be set no less than the period of the longest known repetition frequency in the system.

3.1.5 Sources of uncertainty

EMC measurements are inherently less accurate than most other types of measurement. Whereas, say, temperature or voltage measurement can be refined to an accuracy expressed in parts per million, field strength measurements in particular can be in error by up to 10dB. It is always wise to allow a margin of about this magnitude between your measurements and the specification limits, not only to cover measurement uncertainty but also tolerances arising in production. NAMAS, the body which accredits UK EMC test houses, issues guidelines on determining measurement uncertainty [121] and it requires test houses to report – or at least estimate – their own uncertainties but for EMC tests it does not define acceptable levels of uncertainty. This section discusses how measurement uncertainties arise (Figure 3.14).

3.1.5.1 Instrument and cable errors

Modern self-calibrating test equipment can hold the uncertainty of measurement at the instrument input to within ±1dB. Input attenuator, frequency response, filter bandwidth and reference level parameters all drift with temperature and time and can account for a cumulative error of up to 5dB at the input even of high quality instrumentation. To overcome this a calibrating function is provided. When this is invoked, absolute errors,

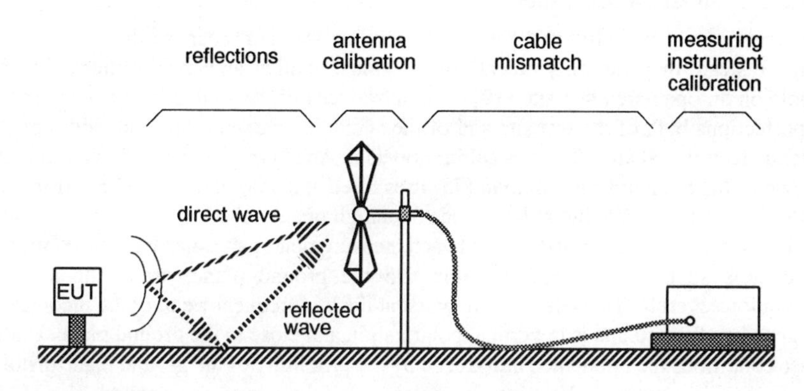

Figure 3.14 Sources of error in radiated emissions tests

switching errors and linearity are measured using an in-built calibration generator and a calibration factor is computed which then corrects the measured and displayed levels. It is left up to the operator when to select calibration, and this should normally be done before each measurement sweep. Do not invoke it until the instrument has warmed up, – typically an hour or so – or calibration will be performed on a "moving target". A good habit is to switch the instruments on first thing in the morning and calibrate them just before use.

The attenuation introduced by the cable to the input of the measuring instrument can be characterised over frequency and for good quality cable is constant and low. The connector can introduce unexpected frequency dependent losses; the conventional BNC connector is particularly poor in this respect and you should perform all accurate measurements with cables terminated in N-type connectors, properly tightened against the mating socket.

Mismatch error

When the cable impedance, nominally 50Ω, is coupled to an impedance that is other than a resistive 50Ω at either end it is said to be mismatched. A mismatched termination will result in reflected signals and the creation of standing waves on the cable. Both the measuring instrument input and the antenna will suffer from a degree of mismatch which varies with frequency and is specified as a Voltage Standing Wave Ratio (VSWR). Appendix C (section C.2) discusses VSWR further. If either the source or the load end of the cable is perfectly matched then no errors are introduced, but otherwise a mismatch error is created which is given by

$$\text{error} = 20 \log_{10} (1 + P_L \cdot P_S)^2 \tag{3.5}$$

where P_L and P_S are the source and load reflection coefficients

As an example, an input VSWR of 1.5:1 and an antenna VSWR of 4:1 gives a mismatch error of ±1dB. The biconical in particular can have a VSWR exceeding 15:1 at the extreme low frequency end of its range. For most measurements, mismatch errors are masked by other sources of error. When the best accuracy is needed, minimize the mismatch error by including an attenuator pad of 6 or 10dB in series with one or both ends of the cable, at the expense of measurement sensitivity.

3.1.5.2 Antenna calibration

Antennas cannot be calibrated against a reference standard because there is no such thing. Instead, they can only be calibrated against another antenna, normally a tuned dipole on an open area test site [19]. This introduces its own uncertainty – due to the imperfections both of the test site and of the standard antenna – into the values of the antenna factors that are offered as calibration data. An alternative method of calibration known as the Standard Site Method [75] uses three antennas and eliminates errors due to the standard antenna, but still depends on a high quality site.

Further, the physical conditions of each measurement, particularly the proximity of conductors such as the antenna cable and the ground plane, affect the antenna calibration severely. These factors are worst at the low frequency end of the biconical's range, when the antenna is in vertical polarization and close to the ground plane. Under these conditions the imbalance introduced by the proximity of the ground plane distorts the antenna's response. Varying the antenna height above the ground plane will introduce a height-related uncertainty in antenna calibration of around 2dB [55].

These problems are less for the log periodic at UHF because nearby objects are normally out of the antenna's near field and do not affect its performance. On the other hand the smaller wavelengths mean that minor physical damage, such as a bent element, has a proportionally greater effect. An overall uncertainty of ±4dB to allow for antenna calibration is not unreasonable.

3.1.5.3 Reflections

The antenna measures not only the direct signal from the EUT but also any signals that are reflected from conducting objects such as the ground plane and the antenna cable. The field vectors from each of these contributions add at the antenna. This can result in an enhancement approaching +6dB or a null which could exceed –20dB. It is for this reason that the height scan referred to in section 3.1.4.3 is carried out; reflections from the ground plane cannot be avoided but nulls can be eliminated by varying the relative distances of the direct and radiated paths. Other objects further away than the defined CISPR ellipse will also add their reflection contribution, which will normally be small (typically less than a dB) because of their distance and presumed low reflection coefficient.

This contribution may become significant if the objects are mobile, for instance people and cars, or if the reflectivity varies, for example trees or building surfaces after a fall of rain. They are also more significant with vertical polarization, since the majority of reflecting objects are predominantly vertically polarized.

Antenna cable

The antenna cable is a primary source of error [54],[55]. By its nature it is a reflector of variable and relatively uncontrolled geometry close to the antenna. There is also a problem caused by secondary reception of common mode currents flowing on the sheath of the cable. Both of these factors are worse with vertical polarization, since the cable invariably hangs down behind the antenna in the vertical plane. They can both be minimized by choking the outside of the cable with ferrite toroid suppressors spaced along it, or by using ferrite loaded RF cable (section 6.1.6.3). If this is not done, measurement errors of up to 5dB can be experienced due to cable movement with vertical polarization.

3.1.5.4 Human and environmental factors

The test engineer

It should be clear from section 3.1.4 that there are many ways to arrange even the simplest EUT to make a set of emissions measurements. Equally, there are many ways in which the measurement equipment can be operated and its results interpreted, even to perform measurements to a well defined standard – and not all standards are well defined. In addition, the quantity being measured is either an RF voltage or an electromagnetic field strength, both of which are unstable and consist of complex waveforms varying erratically in amplitude and time. Although software can be written to automate some aspects of the measurement process, still there is a major burden on the experience and capabilities of the person actually doing the tests.

Some work has been reported which assesses the uncertainty associated with the actual engineer performing radiated emission measurements [74]. Each of four engineers was asked to evaluate the emissions from a desk-top computer consisting of a processor, VDU and keyboard. This remained constant although its disposition was left up to the engineer. The resultant spread of measurement results at various frequencies and for both horizontal and vertical polarization was between 2 and 15dB – which does not engender confidence in their validity! Two areas were recognised as causing this spread, namely differences in EUT and cable configurations, and different exercising methods.

The tests were repeated using the same EUT, test site and test equipment but with the EUT arrangement now specified and with a fixed antenna height. The spread was reduced to between 2 and 9dB, still an unacceptably large range. Further sources of variance were that maximum emissions were found at different EUT orientations, and the exercising routines still had minor differences. The selected measurement time (section 3.1.1.6) can also have an effect on the reading, as can ancillary settings on the test receiver and the orientation of the measurement antenna.

Ambients

The major uncertainty introduced into EMC emissions measurements by the external environment, apart from those discussed above, is due to ambient signals. These are signals from other transmitters or unintentional emitters such as industrial machinery, which mask the signals emitted by the EUT. On an OATS they cannot be avoided, except by initially choosing a site which is far from such sources. In a densely populated country such as the UK this is wishful thinking. A "green-field" site away from industrial areas, apart from access problems, almost invariably falls foul of planning constraints, which do not permit the development of such sites – even if they can be found – for industrial purposes.

Another Catch-22 situation arises with regard to broadcast signals. It is important to be able to measure EUT emissions within the Band II FM and Bands IV and V TV broadcast bands since these are the very services that the emission standards are meant to protect. But the *raison d'être* of the broadcasting authorities is to ensure adequate field strengths for radio reception throughout the country. The BBC publish their requirements for the minimum field strength in each band that is deemed to provide coverage [1] and these are summarized in Table 3.2. In each case, these are (naturally) significantly higher than the limit levels which an EUT is required to meet. In other words, assuming country-wide broadcast coverage is a fact, *nowhere* will it be possible to measure EUT emissions on an OATS at all frequencies throughout the broadcast bands because these emissions will be masked by the broadcast signals themselves. (A novel solution adopted by ICL in Cheshire is to set up their radiated test site underground, in a disused salt mine!)

Service	Frequency range	Minimum acceptable field strength
Long wave	148.5 - 283.5kHz	5mV/m
Medium wave	526.5 - 1606.5kHz	2mV/m
VHF/FM band II	87.5 - 108MHz	54dBμV/m
TV band IV	471.25 - 581.25MHz	64dBμV/m
TV band V	615.25 - 853.25MHz	70dBμV/m

Source: [1]

Table 3.2 Minimum broadcast field strengths in the UK

The only way around the problem of ambients is to perform the tests inside a screened chamber, which is normal practice for conducted measurements but for radiated measurements is subject to severe inaccuracies introduced by reflections from the wall of the chamber as discussed above. An anechoic chamber will reduce these inaccuracies and requirements for anechoic chambers are being introduced into the standards, as mentioned in section 3.1.3.1, but a proper anechoic chamber will be prohibitively expensive for most companies.

Emissions standards such as EN 55 022 recognise the problem of ambient signals and in general require that the test site ambients should not exceed the limits. When they do, the standard allows testing at a closer distance such that the limit level is increased by the ratio of the specified distance to the actual distance. This is usually only practical in areas of low signal strength where the ambients are only a few dB above the limits. Some relief can be gained by orienting the site so that the local transmitters are at right angles to the test range, taking advantage of the antennas' directional response at least with horizontal polarization.

When you are doing diagnostic tests the problem of continuous ambients is less severe because even if they mask some of the emissions, you will know where they are and can tag them on the spectrum display. Some analysis software performs this task automatically. Even so, the presence of a "forest" of signals on a spectrum plot confuses the issue and can be unnerving to the uninitiated. Transient ambients, such as from portable radios or occasional broadband sources, are more troublesome because it is harder to separate them unambiguously from the EUT emissions. Sometimes you will need to perform more than one measurement sweep in order to eliminate all the ambients from the analysis.

Weather

The other environmental factor that affects open area emissions testing, particularly in Northern European climates, is the weather. Some weatherproof but RF-transparent structure is needed to cover the EUT to allow testing to continue in bad weather. The structure can cover the EUT alone, for minimal cost, or can cover the entire test range. Fibreglass and plastics are favourite materials. Wood is not preferred, because the reflection coefficient of some grades of wood is surprisingly high [55]. You may need to make allowance for the increased reflectivity of wet surfaces during and after precipitation.

3.2 RF susceptibility

Until the EMC Directive, most commercial susceptibility testing was not mandatory, but driven by customer requirements for reliability in the presence of interference. Military and aerospace susceptibility test standards have been in existence for some time and have occasionally been called up in commercial contracts in default of any other available or applicable standards. These allow for both conducted and radiated RF susceptibility test methods. The major established commercial standard tests are those listed in IEC 801, which at the time of writing applies radiated field testing only. EN 55 020 requires both conducted and radiated susceptibility tests but applies only to broadcast receivers. Draft standards are in existence which define conducted RF tests for other types of equipment but these are not yet established.

Radiated field susceptibility testing, in common with radiated emissions testing, suffers from considerable variability of results due to the physical conditions of the test set-up. Layout of the EUT and its interconnecting cables affects the RF currents and voltages induced within the EUT to a great extent. At frequencies below the half-wave resonance of the equipment enclosure (see section 4.3.1) cable coupling predominates and hence cable layout and termination must be specified in the test procedure.

3.2.1 Equipment

Figure 3.15 shows the components of a typical radiated susceptibility test set-up in a screened room.

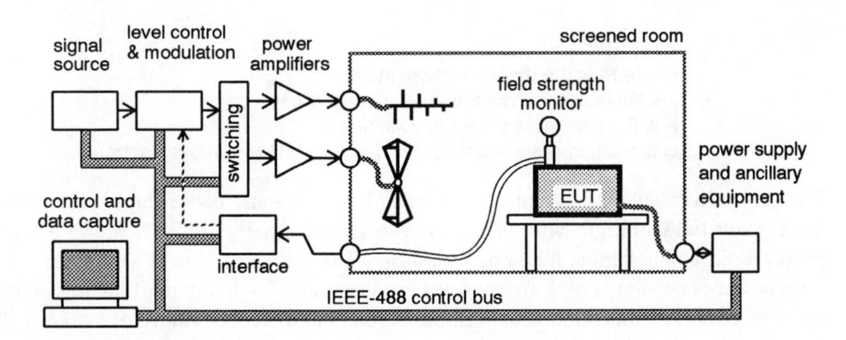

Figure 3.15 RF susceptibility test set-up

The basic requirements are an RF signal source, a broadband power amplifier and a transducer. The latter may be a set of antennas, a transmission line cell or a stripline. These will enable you to generate a field at the EUT, but for accurate control of the field strength there must be some means to monitor it at the EUT, together with control of the level that is fed to the transducer. A test house will normally integrate these components with computer control to automate the frequency sweep and levelling functions.

3.2.1.1 Signal source

Any RF signal generator that covers the required frequency range (27 – 500MHz for IEC801-3, 26 – 1000MHz for draft revisions) will be useable. Its output level must match the input requirement of the power amplifier with a margin of a few dB. This is typically 0dBm and is not a problem.

Draft revisions to IEC801-3 call for the RF carrier to be modulated at 1kHz to a depth of 80%. This can be done within the signal generator or by a separate modulator. The signal generator should either be manually or automatically swept across the output range at 0.005 octaves per second or slower, depending on the speed of response of the EUT, or it can be automatically stepped at this rate in steps of typically 10kHz (but see section 3.2.3.3). The tracking generator output of a spectrum analyser (section 3.1.1.2) is well suited to the former method, or a synthesized signal generator can be used for the latter. The required frequency accuracy depends on whether the EUT exhibits any narrowband responses to interference. A manual frequency setting ability is necessary for when you want to investigate the response around particular frequencies. Be careful that no transient level changes are caused within the signal generator by range changing or frequency stepping.

3.2.1.2 Power amplifier

Most signal sources will not have sufficient output level on their own, and you will require a set of power amplifiers to increase the level. The power output needed will depend on the field strength that you have to generate at the EUT, and on the characteristics of the transducers you use to do this. As well as the antenna factor, an antenna will be characterized for the power needed to provide a given field strength at a set distance. This can be specified either directly or as the gain of the antenna. The relationship between antenna gain, power supplied to the antenna and field strength in the far field is

$$P_t = (r \cdot E)^2/(30 \cdot G) \tag{3.6}$$

where P_t is the antenna power input
r is the distance from the antenna in metres
E is the field strength at r in volts/metre
G is the numerical antenna gain [= antilog(G_{dB}/10)] over isotropic

The gain of a broadband antenna varies with frequency and hence the required power for a given field strength will also vary with frequency. Figure 3.16 shows a typical power requirement versus frequency for a field strength of 10V/m at a distance of 1m. Less power is needed at high frequencies because of the higher gain of the log periodic antenna. Note that the antenna calibration is likely to differ for values of r greater than 1m because of proximity effects.

The power output versus bandwidth is the most important parameter of the power amplifier you will choose and it largely determines the cost of the unit. Very broad band amplifiers (1 – 1000MHz) are available with powers of a few watts, but this may not be enough to generate required field strengths from a biconical antenna in the low VHF region. A higher power amplifier with a bandwidth restricted to 30 – 300MHz will also be needed. If you can use two amplifiers, each matched to the bandwidth and power requirements of the two antennas you are using, this will minimize switching requirements to cover the whole frequency sweep. A typical specification using the antennas of Figure 3.16 to give 10V/m at 1m with a safety margin of 3dB would be 100W below 300MHz, and 3W from 300 to 1000MHz.

Some over-rating of the power output would be desirable to allow for system losses and for the ability to test at a greater distance. If you will be using the system in a non-anechoic screened room (section 3.2.2.1) the system should be over-rated by at least

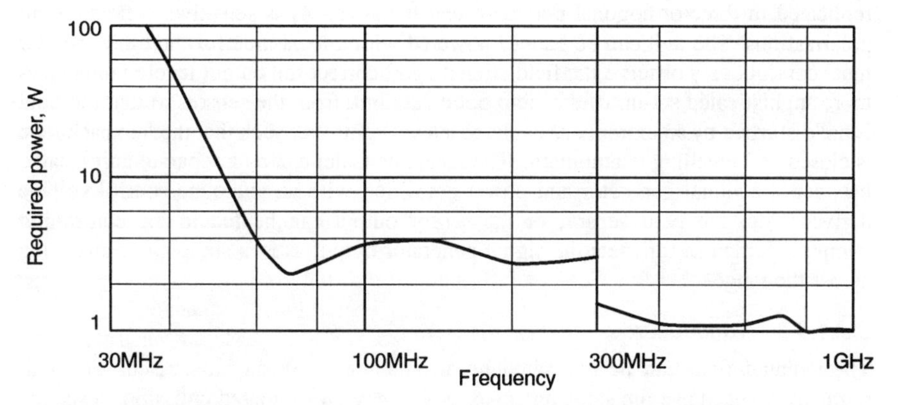

Figure 3.16 Required power versus frequency for 10V/m at 1m, biconical and log periodic antennas

6dB (four times power) to allow for field nulls at certain frequencies due to room reflections. If the system uses other transducers such as a TEM cell or stripline (discussed in section 3.2.1.4) rather than a set of antennas, then the power output requirement for a given field strength will be significantly less. Thus there is a direct cost tradeoff between the type of transducer used and the necessary power of the amplifier.

Other factors that you should take into account (apart from cost) when specifying a power amplifier are:

- linearity: RF susceptibility testing can tolerate some distortion but this should not be excessive, since it will appear as harmonics of the test frequency and may give rise to spurious responses in the EUT; distortion products should be at least –20dB relative to the carrier
- ruggedness: the amplifier should be able to operate at full power continuously, without shutting itself down, into an infinite VSWR, i.e. an open or short circuit load. Test antennas are not perfect, and neither are the working practices of test engineers!
- power gain: full power output must be obtainable from the expected level of input signal, with some safety margin, across the whole frequency band
- reliability and maintainability: in a typical test facility you are unlikely to have access to several amplifiers, so when it goes faulty you need to have assurance that it can be quickly repaired.

3.2.1.3 Field strength monitor and levelling

It is essential to be able to check that the field strength at the EUT is correct. Reflections and field distortion by the EUT will cause different field strength values from those which would be expected in free space, and these values will vary as the frequency band is swept. The only accurate way to maintain the correct level at the EUT is to monitor this continuously with a broadband, isotropic calibrated field sensor. You are recommended to re-read section 3.1.5 on sources of uncertainty in emissions measurements, as the issues discussed there apply equally to measurements of field strength used for susceptibility tests.

The field sensor will normally take the form of a small dipole and detector replicated in three orthogonal planes so that the assembly is sensitive to fields of all polarizations. The unit can be battery powered with a local meter so that the operator must continuously observe the field strength and correct the output level manually. A more sophisticated set-up uses a fibre optic data link from the sensor, so that the field is not disturbed by an extraneous cable, to a level controller such that the feedback loop is closed and levelling is automatic. The level controller can be a separate unit located between the signal generator and power amplifier, with an analogue control voltage derived from the field sensor; or the sensor output can be fed to the controlling computer which in turn sets the signal generator output as it is stepping or sweeping across the range.

3.2.1.4 Transducers

The radiated field can be generated by an antenna as already discussed. You will normally want to use the same antennas as you have for radiated emissions tests, i.e. biconical and log periodic, and this is perfectly acceptable. The power handling ability of these antennas is limited by the balun transformer which is placed at the antenna's feed point. This is a wideband ferrite cored 1:1 transformer which converts the *bal*anced feed of the dipole to the *un*balanced connection of the coax cable (hence bal-un). It is supplied as part of the antenna and the antenna calibration includes a factor to allow for balun losses, which are usually very slight. Nevertheless some of the power delivered to the antenna ends up as heat in the balun core and windings, and this sets a limit to the maximum power the antenna can take.

The high VSWR of broadband antennas (see section 3.1.5.1 and appendix C section C.2), particularly of the biconical at low frequencies, means that much of the feed power is reflected rather than radiated, which accounts for the poor efficiency at these frequencies. Figure 3.17 shows a typical VSWR versus frequency plot for a biconical and a log periodic. As with radiated emissions testing, the plane polarization of the antennas calls for two test runs, once with horizontal and once with vertical polarization.

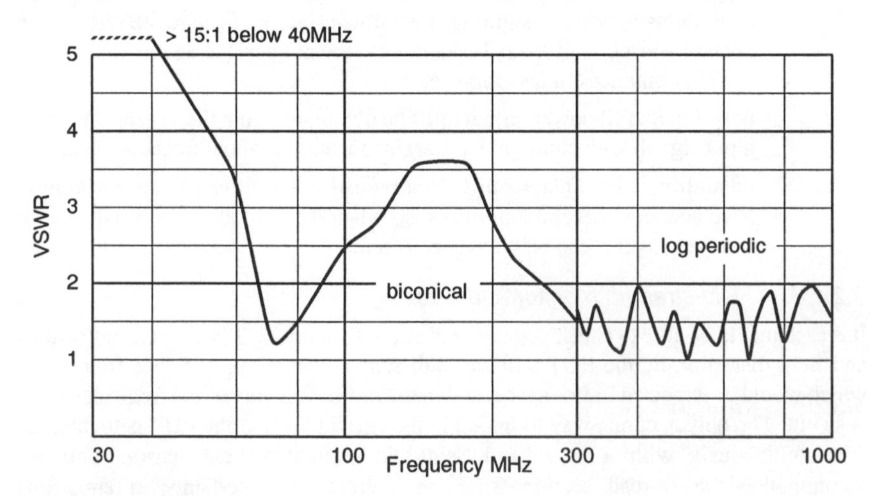

Figure 3.17 VSWR of typical antennas

Two other types of transducer, with superficial similarities, are available for RF susceptibility testing of small EUTs. These are the stripline and the TEM cell (or Crawford cell).

Stripline

The difficulties of testing with antennas led to developments in the 1970s of alternative forms of irradiation of the EUT. Groenveld and de Jong [42] designed a simple transmission line construction which provides a uniform electromagnetic field between its plates over a comparatively small volume, and this has been written in to both IEC 801 part 3 and EN 55 020 as a recommended method of performing part of the radiated susceptibility testing.

The stripline is essentially two parallel plates between which the field is developed, fed at one end through a tapered matching section and terminated at the other through an identical section. The dimensions of the parallel section of line are defined in the standards as 80 x 80 x 80cm, and the EUT is placed within this volume on an insulating support over one of the plates (Figure 3.18). The field between the plates is propagated in TEM (transverse electro-magnetic) mode, which has the same characteristics as free space. The calibration of the stripline is theoretically very simple: assuming proper matching, the field is directly proportional to the voltage at the feed point divided by the distance between the plates,

$$E = V/h \text{ volts per metre} \quad (3.7)$$

In practice some variations from the ideal are likely and calibration using a short probe extending into the test volume is advisable. If the stripline test is conducted in a screened room reflections from the walls will disturb the propagation characteristics quite severely, as they do with antennas, and you will have to surround the stripline with absorbing plates to dampen these reflections. This will be cheaper than lining the walls with anechoic absorber.

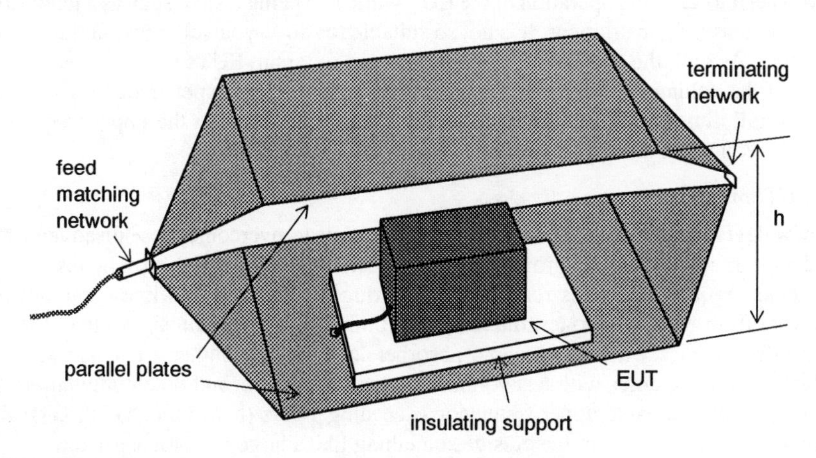

Figure 3.18 The stripline

The accuracy of the stripline depends to a large extent on the dimensions of the EUT. IEC 801-3 recommends that the dimensions should not exceed 25cm, while EN 55 020 allows a height up to 0.7m with a calibration correction factor. Either way, you can only use the stripline on fairly small test objects. There is also an upper frequency restriction of 150 – 200MHz, above which the plate spacing is greater than a half-wavelength and the transmission mode becomes complex so that the field is subject to variability. It would be quite possible though to use the stripline for susceptibility testing below 200MHz (theoretically down to DC if required) along with a log periodic antenna above 200MHz, to get around the unsuitability of the biconical for susceptibility tests below 60MHz. The power requirement of the stripline for a field strength of 10V/m is no more than a couple of watts.

A particular characteristic of testing with the stripline is that the connecting cables for the EUT are led directly through one of the plates and are not exposed to the field for more than a few centimetres. Thus it only tests for direct exposure of the enclosure to the field, and for full susceptibility testing it should be used in conjunction with common mode conducted current or voltage injection. Also, you will need to be able to re-orient the EUT through all three axes to determine the direction of maximum susceptibility.

The TEM cell

An alternative to the stripline for small EUTs and low frequencies is the TEM or Crawford cell. In this device the field is totally enclosed within a transmission line structure, and the EUT is inserted within the transmission line. It is essentially a parallel plate stripline in which one of the plates has been extended to completely enclose the other. Or, you can think of it as a screened enclosure forming one half of the transmission line while an internal plate stretching between the sides forms the other half.

The advantage of the TEM cell, like the stripline, is its small size, low cost and lack of need for high power drive; it can be easily used within the development lab. A further advantage, not shared with the stripline, is that it needs no further screening to attenuate external radiated fields. The disadvantage is that a window is needed in the enclosure if you need to view the operation of the EUT while it is being tested, such as a television set or a measuring instrument. It is not so suitable for do-it-yourself construction as the stripline. As with the stripline, it can only be used for small EUTs (dimensions up to a third of the volume within the cell) and it suffers from a low upper frequency limit. If the overall dimensions are increased to allow larger EUTs, then the upper frequency limit is reduced in direct proportion.

The GTEM

A new development, the GTEM cell [37][43], claims to overcome these disadvantages and appears to hold out the promise of lower cost, well defined testing. The restriction on upper frequency limit is removed by tapering the transmission line continuously outward from feed point to termination, and combining a tapered resistive load for the lower frequencies with an anechoic absorber load for the higher frequencies. This allows even large units, with test volume heights up to 1.75m and potentially larger, to be made with a useable upper frequency exceeding 1GHz (hence the "G" in GTEM). The actual unit looks from the outside something like a large horn loudspeaker.

The GTEM has clear advantages for immunity testing – it allows the full frequency range to be applied in one sweep, without the need for a screened enclosure or high power amplifiers – and it could potentially be used for radiated emissions testing also, if the results obtained from it can be correlated with those from an open field site. This may turn out to be difficult to achieve because of the effects of cable layout and the differing polarizations of the two methods. Its advantages are so attractive though, particularly in terms of allowing one relatively inexpensive (c. £50,000) facility to perform all RF EMC testing, that considerable resources are being put into characterizing the GTEM's operation and in persuading the standards authorities to accept it as an alternative test method.

3.2.2 Facilities

RF susceptibility testing, like radiated emissions testing, cannot be readily carried out on the development bench. You will need to have a dedicated area set aside for these tests – which may be in the same area as for the emissions tests – which includes the RF field generating equipment and, most importantly, has a screened room.

3.2.2.1 The screened room

RF susceptibility tests covering the whole frequency bands specified in the standards should be carried out in a screened room to comply with various national regulations prohibiting interference to radio services. Recommended shielding performance is at least 100dB attenuation over the range 10MHz to 1GHz [97]; this will reduce internal field strengths of 10V/m to less than 40dBμV/m outside. The shielding attenuation depends on the constructional methods of the room in exactly the same way as described for shielded equipment enclosures in section 6.3. It is quite often possible to trade off performance against reduced construction cost, but a typical high-performance room will be built up from modular steel-and-wood sandwich panels, welded or clamped together. Ventilation apertures will use honeycomb panels; the room will be windowless. All electrical services entering the chamber will be filtered. Lighting will be by incandescent lamps as fluorescent types emit broadband interference. The access door construction is critical, and it is normal to have a double wiping action "knife-edge" door making contact all round the frame via beryllium copper finger strip.

In addition, the screened room isolates the test and support instrumentation from the RF field. The interconnecting cables leaving the room should be suitably screened and filtered themselves. A removable bulkhead panel is often provided which can carry interchangeable RF connectors and filtered power and signal connectors. As well as for RF susceptibility tests, a screened room is useful for other EMC tests as it establishes a good ground reference plane and an electro-magnetically quiet zone. Figure 3.19 shows the features of a typical screened chamber installation.

Room resonances

An un-lined room will exhibit field peaks and nulls at various frequencies determined by its dimensions. The larger the room, the lower the resonant frequencies. This phenomenon is discussed again in section 4.3.1 and equation (4.6) gives the lowest resonant frequency. For a typical room of 2.5 x 2.5 x 5m this works out to around 70MHz.

To damp these resonances the room can be lined with absorber material, typically carbon loaded foam shaped into pyramidal sections, which reduces wall reflections. The room is then said to be "anechoic" if all walls and floor are lined or "semi-anechoic" if not. Such material is expensive – a fully-lined room will be more than double the cost of an unlined one – and is in any case ineffective below about 200MHz. Very large absorber pyramids are needed for lower frequencies and these reduce the useable volume of the room unacceptably. Measurements made in the frequency range 30–200MHz are therefore subject to very large uncertainties, of the order of 40dB, at the resonant frequencies, and are also not repeatable, because a small change in the

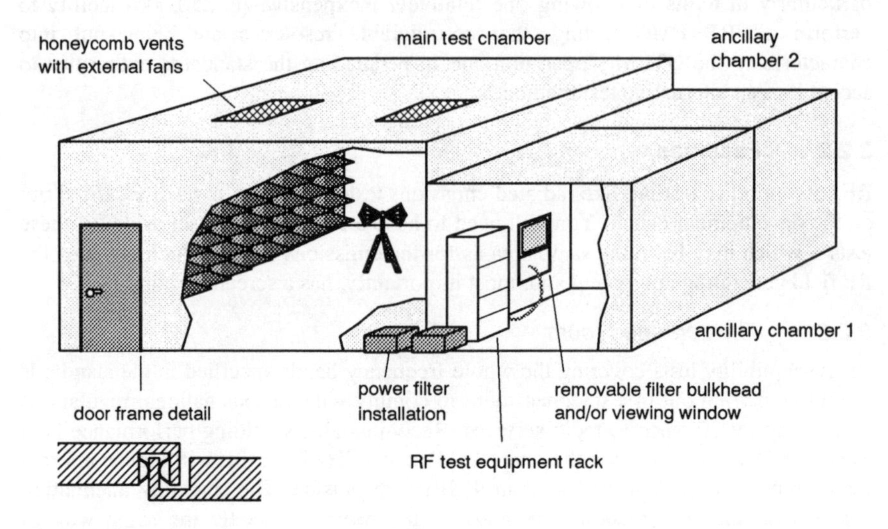

Figure 3.19 Typical screened room installation

antenna or EUT position can give a large change in the field distribution. Figure 3.20 illustrates the frequency response of a typical un-lined empty room.

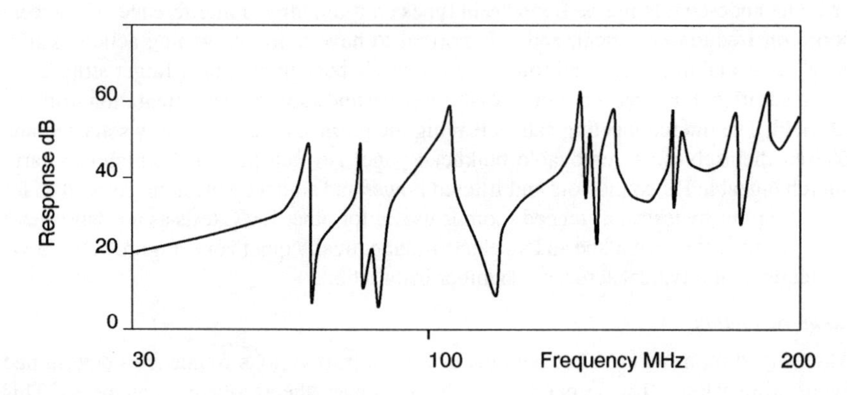

Figure 3.20 Unlined screened room frequency response

An alternative to pyramidal absorbers is to line the walls with ferrite tiles or ferrite grid absorbers. Some Japanese suppliers of these materials claim extremely good results in damping room resonances, but the ferrites are also expensive and bring their own problems in fixing and mechanical support. Careful placement of absorber blocks at voltage maxima within the room, or of ferrite absorbers at current maxima on the walls, has been shown [20][33] to have a significant effect on resonances at comparatively little cost.

3.2.2.2 Ancillary equipment

You will need a range of support equipment in addition to the RF test equipment described in detail in section 3.2.1. Obviously, control and data capture computing equipment will be required for a comprehensive set-up. Various test jigs and coupling networks, depending on the type of EUT and the detail of the standards in use, must be included. Beyond that, some form of communication will be needed between the inside of the screened room and the outside world. This could take the form of RFI-proof CCTV equipment, intercoms or fibre optic data communication links.

The ancillary equipment housed outside the screened room will also include all the support equipment for the EUT. Test houses will normally have two subsidiary screened chambers abutting the main one, one of which contains the RF test instrumentation, the other housing the support equipment. This ensures that there is no interaction between the external environment, the RF instrumentation and the support equipment. Provided the environment is not too noisy and the RF instrumentation is individually well screened, you do not really need these two extra screened chambers for your own EMC testing.

3.2.3 Test methods

As with radiated emissions, the major concern of standardized susceptibility test methods is to ensure repeatability of measurements. The susceptibility test is complicated by not having a defined threshold which indicates pass or failure. Instead, a (hopefully) well defined level of interference is applied to the EUT and its response is noted. The test procedure concentrates on ensuring that the applied level is as consistent as possible and that the means of application is also consistent.

3.2.3.1 Preliminary checking

You will need to carry out some preliminary tests to find the most susceptible configuration and operating mode of the EUT. If it is expected to pass the compliance test with a comfortable margin, you may need to apply considerably greater field strengths in order to deliberately induce a malfunction. Hopefully, with the initially defined set-up and operation there will be some frequency and level at which the operation is corrupted. This is easier to find if the EUT has some analogue functions, which are perhaps affected to a small degree, than if it is entirely digital and continues operating perfectly up to a well-defined threshold beyond which it crashes completely.

Once a sensitive point has been found, you can vary the orientation, cable layout, grounding regime and antenna polarization to find the lowest level which induces a malfunction at that frequency. Similarly, the operating mode can be changed to find the most sensitive mode. It is often worthwhile incorporating special test software to continuously exercise the most sensitive mode, if this is not part of the normal continuous operation of the instrument. Note that some changes may do no more than shift the sensitive point to a different frequency, so you should always repeat a complete frequency sweep after any fine tuning at a particular frequency.

3.2.3.2 Compliance tests

Once the sensitive configuration has been established it should be carefully defined and rigorously maintained throughout the compliance test. Changes in configuration halfway through will invalidate the testing. If there are several sensitive configurations these should be fully tested one after the other. Notwithstanding this, equipment should always be tested in conditions that are as close as possible to a typical installation – that is with wiring and cabling as per normal practice, and with hatches and covers in place. If the wiring practice is unspecified, leave a nominal length of 1m of cable exposed to the incident field. If the EUT is a rack or cabinet it will be placed on but insulated from the floor, otherwise on a wooden table. The antennas will normally be placed at least 1m from it, at a greater distance if possible consistent with generating an adequate field strength.

The compliance test will concentrate on making sure that the specified test level is maintained throughout the frequency sweep. This will be achieved by using the field strength monitoring equipment referred to in section 3.2.1.3, or by using a monitoring probe for conducted interference injection. The parameters which have been chosen to represent the operation of the EUT must be continuously monitored throughout the sweep, preferably by linking them to an automatic data capture and analysis system.

Assuming that the EUT remains correctly operational throughout the sweep, i.e. it passes, it can be useful to know how much margin there is in hand at the sensitive point(s). You can do this by repeating the sweep at successively higher levels and mapping the EUT's response. This will indicate both the margin you can allow for production variability, and the possibilities for cost reduction by removing suppression components.

3.2.3.3 *Sweep rate*

The sweep rate itself may be critical to the performance of the EUT. For many systems there may be little sensitivity to sweep rate since demodulation of applied RF tends to have a fairly broad bandwidth; usually, responses are caused by structural or coupling resonances which are low-Q and therefore several MHz wide. On the other hand, some frequency sensitive functions in the EUT may have a very narrow detection bandwidth so that responses are only noted at specific frequencies. This may easily be the case, for instance, with analogue-to-digital converters operating at a fixed clock frequency, near which interfering frequencies are aliased down to the baseband. If the sweep rate through these frequencies is too fast (or the step spacing is too great) then a response may be missed. Such narrowband susceptibility may be 25–30dB worse than the broadband response. Therefore some knowledge of the EUT's internal functions is essential, or considerably more complex test procedures are needed.

3.2.3.4 *Safety precautions*

At field strengths not much in excess of those defined in many susceptibility standards, there is the possibility of a biological hazard from the RF field arising to the operators if they remain in the irradiated area for an appreciable time. For this reason a prudent test facility will not allow its test personnel inside a screened chamber while a test is in progress, making it necessary for a remote monitoring device (such as a CCTV) to be installed for some types of EUT.

Health and safety legislation differs between countries. In the UK at the time of writing there are no *mandatory* requirements placed on maximum permissible RF field exposure, but the National Radiological Protection Board (NRPB) has published guidelines [122] for this purpose and it would be advisable from the point of view of potential health and safety claims for any employer to adhere to these. The guidelines are based on (but not identical to) guidance published by the International Non-Ionizing Radiation Committee [123], and take into account the known thermal and electric shock effect of RF fields. They do not consider possible athermal effects, which is a highly controversial field of study and for which no firm guidance has yet been produced. The guidelines contained in [122] for occupational exposure to continuous fields over the frequency range of interest for RF susceptibility testing are reproduced in Table 3.3.

Frequency range	RMS field strength
30 to 400MHz	61.4 V/m
400 to 2000MHz	97.1 · √F (GHz) V/m

Table 3.3 NRPB guidelines for maximum field strength exposure

3.2.3.5 Short cuts in susceptibility testing

There will be many firms which decide that they cannot afford the expense of a full RF susceptibility set up, including a screened room, as described in section 3.2.2. One possibility is to restrict the test frequencies to the "free radiation" frequencies as permitted by international convention, on which unrestricted emissions are allowed. These are primarily intended for the operation of industrial, scientific and medical equipment and are listed in Table 1.1 on page 5. Another course known to be taken by some firms, is to use the services of a licensed radio amateur transmitting on the various amateur bands available to them – 30MHz, 50MHz, 70MHz, 144MHz and 432MHz; although this is strictly outside the terms of the amateur radio license.

In either case the use of particular frequencies removes the need for a screened room to avoid interference with other services. If the EUT's response to RF interference was broadband across the whole frequency range then spot frequency testing would be adequate, but this is rarely so; resonances in the coupling paths emphasize some frequencies at the expense of others, even if the circuit response is itself broadband. It is therefore quite possible to believe an optimistic performance of the EUT if you have only tested it at discrete frequencies, since resonant peaks may fall between these. A compliance test must always cover the entire range.

Transient testing

In practice, it has been found that for many digital products transient and ESD performance is linked to good RF immunity, since responsive digital circuits tend to be sensitive to both phenomena. Therefore much development work can proceed on the basis of transient tests, which are easier and less time-consuming to apply than RF tests, and are inherently broadband. Where analogue circuits are concerned then a proper RF field test is always necessary, since the demodulated offset voltage which RF injection causes cannot be simulated by a transient. But a minimal set of transient plus spot frequency RF tests may give you an adequate assessment of the product's immunity during the development stages.

3.2.4 Conducted RF susceptibility

As we have noted, the majority of commercial standards call for radiated field susceptibility tests. But because of the difficulty and expense of performing these tests, there is growing pressure to allow conducted susceptibility testing at the lower frequencies, by analogy to the distinction between radiated and conducted emissions testing. The draft standards EN 55 101 part 4 and IEC 801 part 6 both propose conducted susceptibility test methods. Although these are (at the time of writing) different from each other, they both require the interfering signal to be injected in common mode on each test port. The immunity standard for broadcast receivers,

EN 55 020, also defines test methods for immunity from conducted RF currents and voltages. The method of "bulk current injection" (BCI) developed within the aerospace and military industries for testing components of aircraft systems, and now being adapted for application to automotive components, is another similar technique. See also section 4.3.1.2.

3.2.4.1 Coupling methods

Any method of cable RF injection testing requires that the common mode impedance at the end of the cable remote from the EUT is defined. Thus each type of cable must have a common mode decoupling network at its far end, to ensure this impedance and to isolate any ancillary equipment from the effects of the RF current on the cable. (This is exactly analogous to the mains LISN used for emission testing and discussed in section 3.1.2.2.) Voltage injection in addition requires that this network is used to couple the RF voltage onto the cable, a complication which is absent when current injection via a clamp-on probe is used. Thus a test house which handles these methods must have a wide range of coupling/decoupling networks available, to cater for the variety of different cable and signal types that will come its way. If your company makes equipment which predominantly uses only one or two types of cable – say single-channel RS-232 data links and mains – then this is not an onerous requirement. Figure 3.21 shows the general arrangement for making conducted susceptibility tests.

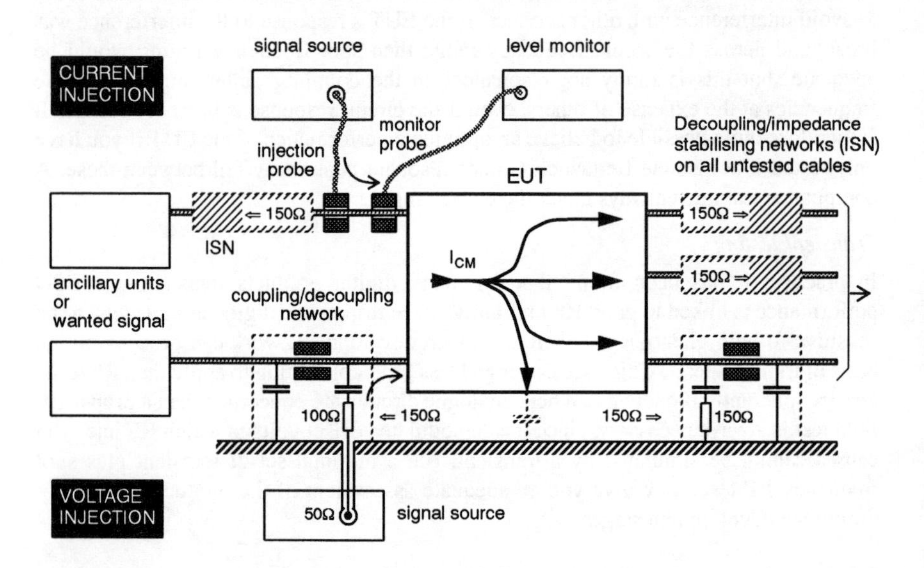

Figure 3.21 Conducted susceptibility test set-up

3.2.4.2 Disadvantages and restrictions

Conducted susceptibility testing has the major advantage of not requiring expensive screened room facilities, but it does have some disadvantages. It is particularly questionable whether it accurately represents real situations when there are several cables connected to the EUT. When the whole system is irradiated then all cables would be carrying RF currents, but in most conducted susceptibility test methods only one is tested at a time. Each of the other cables represents a common mode load on the test system and this must be artificially created by including extra impedance stabilising networks on them. Networks for direct coupling to cables with many signal lines are expensive to construct, bulky and may adversely affect the signal line characteristics. The method is therefore less well suited to equipment which has many cables connected to it routinely.

The current probe method has the considerable advantage, compared to voltage injection, that it is non-intrusive; the probe just clamps over the cable to induce a common mode current with no direct connection to it. This makes it very attractive for cables with many conductors. It has the disadvantage, though, that the stray capacitance between probe and cable is ill defined. This restricts its useable frequency range since at the higher frequencies, both the inductive coupling path and the cable common mode impedance are seriously affected by this capacitance.

A further question hangs over the levels used to establish compliance. When either voltage or current injection are used, the actual power applied to the EUT will depend on the common mode impedance at the EUT port: a low impedance will run the risk of over-testing if a voltage source is used, and under-testing if a current source is used, and vice versa for a high impedance. To get around this problem, some standards specify limits in terms of current into a calibration jig, for which the EUT is then substituted at a constant power level.

Frequency range

The major restriction on conducted susceptibility testing is one of frequency. For EUT sizes much less than the wavelength of the test frequency, the dominant part of the RF energy passing through equipment that is exposed to a radiated field is captured by the cables, and therefore conducted testing is representative of reality. As the frequency rises so that the EUT dimensions approach a half-wavelength, the dominance of the cable route reduces and at higher frequencies the field coupling path is a complex and unpredictable mixture of interactions with the EUT structure and internal circuits, as well as with its cables. For this reason the upper frequency limit is restricted in existing and draft standards to between 80 and 230MHz (corresponding to equipment dimensions of between about 0.6m and 2m). For higher frequencies, radiated testing is still necessary.

3.3 ESD and transient susceptibility

By contrast with RF testing, ESD and transient test methods are rather less complicated and need less in the way of sophisticated test equipment and facilities. Nevertheless the bandwidth of fast transients and of the electrostatic discharge is very wide and extends into the VHF region, so many precautions that are necessary for RF work must also be taken when performing transient tests.

3.3.1 ESD

3.3.1.1 Equipment

The electrostatic discharge generator described in IEC 801 part 2 is fairly simple. The circuit is shown in Figure 3.22. The main storage capacitor C_s is charged from the high voltage power supply via R_{ch} and discharged to the EUT via R_d and the discharge switch. The switch is typically a vacuum relay under the control of the operator.

Compliance testing uses single discharges, but for exploratory testing the capability of a fast discharge rate of 20 per second is suggested. The output voltage should reach 8kV for contact discharge, or 15kV if air discharge is included, although for the tests required in the present provisional immunity standards lower voltages are specified.

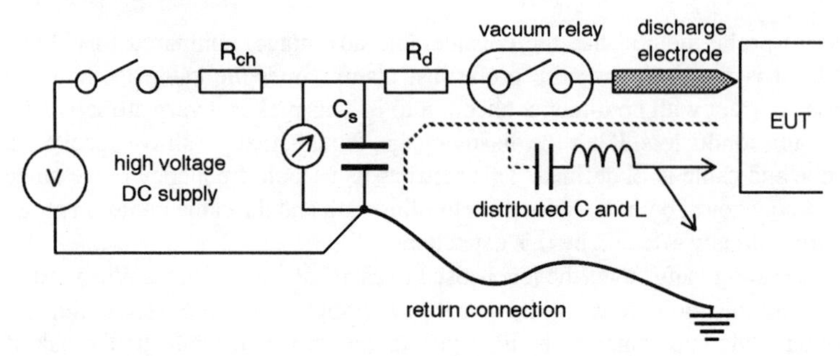

Figure 3.22 ESD generator (per IEC 801 part 2)

The critical aspect of the ESD generator is that it must provide a well defined discharge waveform with a rise time of between 0.7 and 1 nanosecond. This implies that the construction of the circuit around the discharge electrode is important; C_s, R_d and the discharge switch must be placed as close as possible to the discharge electrode which itself has specified dimensions. A round tip is used for air discharge, and a sharp tip for contact. The distributed capacitance and inductance of the electrode and associated components forms part of the discharge circuit and essentially determines the initial rise time, since the return connection to the EUT is relatively long (2m) and its inductance blocks the initial discharge current. As these distributed parameters cannot be satisfactorily specified, the standard requires that the generator's waveform is calibrated in a special test jig using an oscilloscope with a bandwidth of at least 1GHz.

If you use a ready-built ESD generator this calibration will have already been done by the manufacturer, though it should be re-checked at regular intervals. If you build it yourself you will also have to build and use the calibration jig.

3.3.1.2 Test methods

Because of the very fast edges associated with the ESD event, high frequency techniques are essential in ESD testing. The use of a ground reference plane is mandatory; this can of course be the floor of a screened room, or the same ground plane that you have installed for the tests outlined in section 3.1.3. You may want to apply ESD tests to equipment after it has been installed in its operating environment, in which case a temporary ground plane connected to the protective earth should be laid near to the equipment. Other co-located equipment may be adversely affected by the test, so it is wise not to carry out such tests on a "live" operating system.

For laboratory tests, the EUT should be set up in its operating configuration with all cables connected and laid out as in a typical installation. The connection to the ground is particularly important, and this should again be representative of installation or user practice. Table-top equipment should be placed on a wooden table 80cm over the ground plane, with a horizontal coupling plane directly underneath it but insulated from it. Floor standing equipment should be isolated from the ground plane by an insulating support of about 10cm. Figure 3.23 illustrates a typical set-up. Any ancillary equipment should itself be immune to coupled ESD transients, which may be induced from the field generated by the ESD source/EUT system or be conducted along the connected cables.

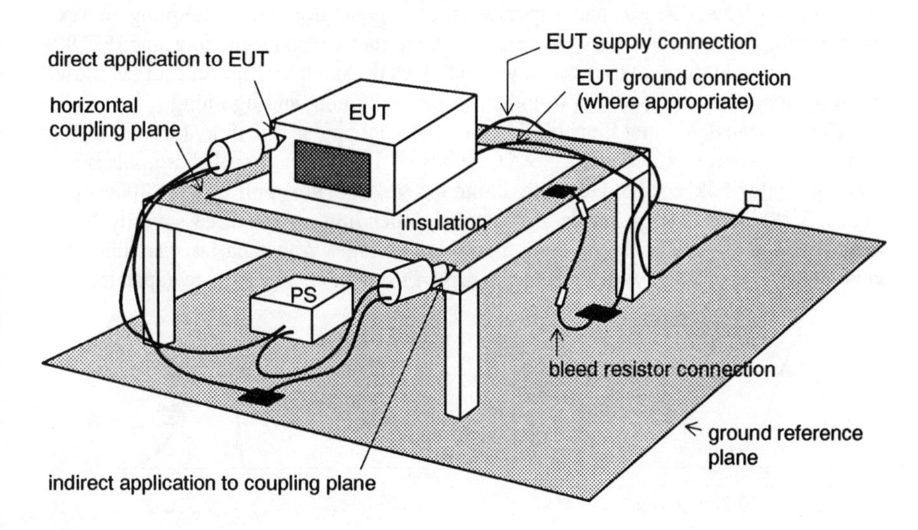

Figure 3.23 ESD test set-up

Discharge application

The actual points of application of the discharges should be selected on the basis of exploratory testing, to attempt to discover sensitive positions. These points should only be those which are accessible to personnel in normal use. For the exploratory testing, use a fast repetition rate and increase the applied voltage from the minimum up to the specified test severity level, to ascertain any threshold of malfunction. Also, select both polarities of test voltage. Compliance testing requires the specified number of single discharges with at least 1 second between them, on each chosen point at the specified test level and with the most sensitive polarity, or with both polarities.

Contact discharge is preferred, but this requires that the EUT has conducting surfaces or painted surfaces which are not regarded as insulating. For a product where this is not possible (e.g. with an overall plastic enclosure) use air discharge to investigate user accessible points where breakdown to the internal circuit might occur, such as the edges of keys or connector or ventilation openings.

In cases where neither direct contact nor air discharge application to the EUT is possible, and to simulate discharges to objects near to the equipment in its operating environment, the discharge is applied to a coupling plane located a fixed distance away from the EUT. This can be either the horizontal coupling plane shown in Figure 3.23 or an ancillary vertical coupling plane.

3.3.2 Transients

3.3.2.1 Equipment

When testing equipment for susceptibility to conducted transients the transients themselves, and the coupling network by which the transients are fed into the ports must be well defined. The network must decouple the side of the line furthest from the EUT and at the same time provide a fixed impedance for the coupling route. In this respect it is similar to the LISN used in emissions testing, and the CDNs used for conducted RF susceptibility tests. IEC 801 part 4 specifies the test generator and the coupling methods for bursts of fast transients such as are caused by inductive load switching, and IEC 801 part 5 (still in draft form at the time of writing) does the same for transient surges caused by overvoltages or overcurrents from heavy load switching and lightning.

The fast transient burst is specified to have a single pulse rise time/duration of 5ns/50ns from a source impedance of 50Ω. Bursts of 15ms duration of these pulses at a repetition rate of 5kHz (2.5kHz at maximum test voltage) are applied every 300ms (see Figure 3.24). The voltage levels are selected depending on specified severity levels from 250V to 4kV. In order to obtain these high voltages with such fast rise times, the generator is constructed with a spark gap driven from an energy storage capacitor.

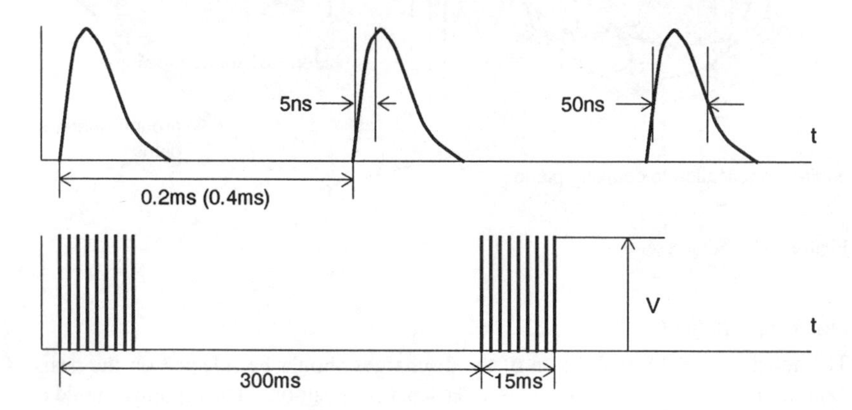

Figure 3.24 Fast transient burst specification (per IEC 801 part 4)

The coupling network for power supply lines applies the pulse in common mode to each line via an array of coupling capacitors, while the source of each line is also decoupled by an LC network. Coupling onto signal lines uses a capacitive clamp, essentially two metal plates which sandwich the line under test to provide a distributed coupling capacitance and which are connected to the transient generator. Any associated equipment which may face the coupled transients must obviously be immune to them itself.

3.3.2.2 Test methods

As with ESD tests, a reference ground plane must be used. This is connected to the protective earth on the decoupled side of the transient coupling network. Floor standing equipment is stood off from this ground plane by a 10cm insulating block, and table top EUTs are placed on an insulating table 80cm above it. A 1m length of mains cable connects the EUT to the coupling network, which itself is bonded to the ground plane. If the EUT enclosure has a separate protective earth terminal, this is connected to the ground plane via the coupling network and transients are applied directly to it also. I/O cables are fed through the capacitive clamp which is located 10cm above the ground plane. A typical set up is shown in Figure 3.25.

Actual application of the transients is relatively simple, compared to other

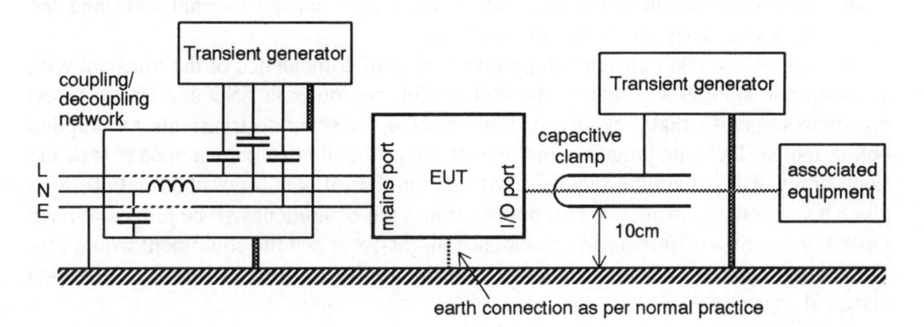

Figure 3.25 Fast transient test set-up

susceptibility tests. No exploratory testing is necessary except to determine the most sensitive operating mode of the equipment. Typically, bursts are applied for a duration of 1 minute in each polarity on each line to be tested. The required voltage levels are defined in the relevant standard, and vary depending on the anticipated operating environment and on the type of line being tested.

Most surge protectors have low average power capabilities even though their peak power dissipation may be high. The repetition rate of surge application when you apply surge voltages to protected inputs may be limited by this factor. When surges are applied to the mains input, they should be synchronised with the mains waveform so as to occur at the worst case point on it (normally the positive and negative peaks).

3.3.3 Sources of uncertainty

Assuming that the EUT's response can be accurately characterized, the major uncertainties in transient testing stem from repeatability of layout and the statistical nature of the transient application. The climatic conditions may also have some bearing on the results of air discharge ESD tests.

3.3.3.1 Layout

The wide bandwidth of the ESD and fast burst transients means that cables and the EUT structure can act as incidental radiators and receptors just as they do in RF testing. Refer to section 3.1.5 in this connection. Therefore the test layout, and routing and termination of cables, must be rigorously defined in the test plan and adhered to throughout the test. Variability will affect the coupling of the interference signals into and within the EUT and may to a lesser extent affect the stray impedances and hence voltage levels. Equally, variability in the EUT's build state, such as whether metal panels are in place and tightened down, will have a major effect on ESD response.

3.3.3.2 Transient timing

In a digital product the operation is a sequence of discrete states. When the applied transient is of the same order of duration as the states (or clock period), as is the case with the ESD and fast burst transients, then the timing of application of the transient with respect to the internal state will affect the unit's susceptibility. If the pulse coincides with a clock transition then the susceptibility is likely to be higher than during a stable clock period. There may also be some states when the internal software is more immune than at other times, for example when an edge triggered interrupt is disabled. Under most circumstances the time relationship between the internal state and the applied transient is asynchronous and random.

Therefore, for fast transients the probability P of coincidence of the transient with a susceptible state is less than unity, and for this reason both ESD and transient test procedures specify that a relatively large number of separate transients are applied before the EUT can be judged compliant. If the probability of coincidence P is of the same order or less than the reciprocal of this number, it is still possible that during a given test run the coincidence will not occur and the equipment will be judged to have passed, when on a different run coincidence might occur and the equipment would fail. There is no way around this problem except by applying more test transients in such marginal cases.

3.3.3.3 Environment

In general, the non-electromagnetic environmental conditions do not influence the coupling of interference into or out of electronic equipment, although they may affect the operational parameters of the equipment itself and hence its susceptibility. The one exception to this is with air discharge ESD. In this case, the discharge waveform is heavily influenced by the physical orientation of the discharge electrode and the rate of approach to the EUT, and also by the relative humidity of the test environment. This means that the test repeatability will vary from day to day and even from hour to hour, all other factors being constant, and is the main reason why the air discharge method has fallen out of favour.

3.4 Evaluation of results

The variety and diversity of equipment and systems makes it difficult to lay down general criteria for evaluating the effects of interference on electronic products. Nevertheless, the test results can be classified on the basis of operating conditions and the functional specifications of the EUT according to the criteria discussed below.

It is up to the manufacturer to specify the limits which define "degradation or loss of function", and to decide which of these criteria should be applied to each test. Such specifications may be prompted by preliminary testing or by known customer requirements. In any case it is important that they are laid out in the final EMC test plan for the equipment. If the equipment is being supplied to a customer on a one-to-one contractual basis then clearly there is room for mutual agreement and negotiation on acceptance criteria, but this is not possible for products placed on the mass market, which have only to meet the essential requirements of the EMC Directive. In these cases, you have to look to the immunity standards for general guidance.

3.4.1 Performance criteria

The generic immunity standard [94] lays down guidelines for criteria against which to judge the EUT's performance when the various test levels are applied. These are quoted in full in Appendix B (section B.2.4) and also in section 1.3.6.2, but can be summarized as follows:

> *Performance criterion A: The apparatus shall continue to operate as intended. No degradation of performance or loss of function is allowed below a performance level or a permissible loss of performance specified by the*

manufacturer.

This criterion applies to phenomena which are normally continuously present, such as RF interference.

Performance criterion B: *The apparatus shall continue to operate as intended after the test. No degradation of performance or loss of function is allowed below a performance level specified by the manufacturer, when the apparatus is used as intended. During the test, degradation of performance is however allowed. No change of actual operating state or stored data is allowed.*

This applies to transient phenomena.

Performance criterion C: *Temporary loss of function is allowed, provided the loss of function is self recoverable or can be restored by the operation of the controls.*

This applies to mains interruption.

If you do not specify the minimum performance level or the permissible performance loss, then either of these may be derived from the product description and documentation (including leaflets and advertising) and what the user may reasonably expect from the apparatus if used as intended. Thus, for example, if a measuring instrument has a quoted accuracy of 1% under normal conditions, it would be reasonable to expect this accuracy to be maintained when subject to RF interference at the level specified in the standard, unless your operating manual and sales literature specifies a lower accuracy under such conditions. It may lose accuracy when transients are applied, but must recover it afterwards. A personal computer may exhibit distortion or "snow" on the image displayed on its video monitor under transient interference, but it must not crash nor suffer corruption of data.

Product specific criteria

Some product specific immunity standards may be able to be more precise in their definition of acceptable performance levels. For example EN 55 020, applying to broadcast receivers, specifies a wanted to unwanted audio signal ratio for sound interference, and a just perceptible picture degradation for vision interference. Even this relatively tight definition may be open to interpretation. Another possibility for telecommunications equipment is to comply with a defined criterion for bit error rate and loss of frame alignment.

The subjectivity of immunity performance criteria will present a major headache when the full legal implications of the EMC Directive come to be tested. It will undoubtedly be open to manufacturers to argue when challenged, not only that differing test procedures and layouts have been used in judging compliance of the same product, but that differing criteria for failure have also been applied. It will be in their own interest to be clear and precise in their supporting documentation as to what they believe the acceptance criteria are, and to ensure that these are in line as far as possible with the above generic guidelines. Even so, the unsatisfactorily vague nature of these guidelines will act as a spur to the development of many product specific immunity standards in the coming years.

3.5 Mains harmonic emission

Harmonic components of the AC supply input current to an item of equipment arise from non-linearities of the load over a single cycle of the input voltage. The EMC Directive will include requirements for measuring harmonic emissions as embodied in the standard EN 60 555 part 2 [92]. Mains harmonics are also discussed further in section 4.2.3.

Although the harmonic frequency range under consideration extends only up to 2kHz (the 40th harmonic of 50Hz), and therefore does not by any stretch of the imagination need to employ RF measurement techniques, there are still some aspects of the measurement which may not be entirely obvious and should be considered further [45].

3.5.1 Equipment

EN 60 555 part 2 defines the method of measurement and each item of test equipment is specified. Figure 3.26 shows the basic measurement circuit, and its components are

- an AC source;
- a current transducer;
- a wave analyser.

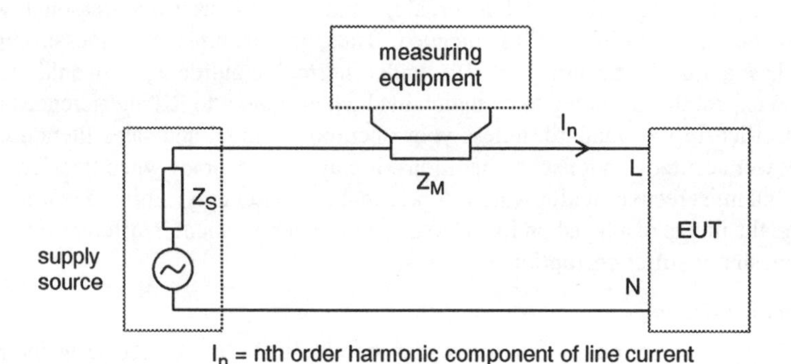

Figure 3.26 Mains harmonic emission measurement circuit

3.5.1.1 AC supply source

To make a harmonic measurement with the required accuracy you need a source with very low distortion, high voltage stability and settability and low impedance. In general the public mains supply will not be able to meet these requirements. The voltage must be adjustable over the range ±6% from a nominal of 220–240V, and be stable to within ±2% of the selected level during the measurement. The harmonic distortion must be less than 0.4% at third harmonic, 0.2% at 5th, 0.15% at 7th and 0.1% at all others – at present this applies only to TV receiver measurements, but will cover measurements of most other electronic equipment when the scope of the standard is extended. The source impedance (including that of the transducer and measuring equipment) must not change the load current from the ideal value, i.e. that measured with a zero source impedance, at any harmonic frequency by more than 5% of the permissible limits.

To meet these requirements you will normally use a power amplifier driven by a 50Hz sine-wave oscillator, with negative feedback to maintain the low output impedance. The ouptut may be fed through a power transformer for voltage step-up purposes, but the transformer reactance must not be allowed to affect the output impedance at the higher harmonic frequencies. Variacs are not recommended for the same reason. The amplifier will need to be large to cope with the full range of loads – the standard covers equipment rated up to 16A, which is a power level of 4000W, although for in-house use your product range may not approach this level and a smaller amplifier would suffice. For high power and highly distorting loads the "model" AC source becomes quite difficult to realise.

3.5.1.2 Current transducer

The current transducer couples the harmonic current I_n to the measuring instrument, and it can be either a current shunt or a current transformer. In both cases, the transducer impedance Z_M is added to the source output impedance and the two together must cause negligible variation in the load current harmonic structure. A shunt of less than 0.1Ω impedance and a time constant less than 10μs is acceptable, but does not provide any isolation from the measuring circuit. A current transformer does offer isolation, but will need to be calibrated at each harmonic frequency and may suffer from saturation when the measured current includes a DC component.

3.5.1.3 Wave analyser

The wave analyser measures the amplitude of each harmonic component I_n for n = 2 to 40. It can be either a frequency domain type, using selective filters or a spectrum analyser, or a time domain type using digital computation to derive the discrete Fourier transform. The error in measuring a constant value must be less than 5% of the permissible limit or 0.015A, whichever is greater. EN 60 555 part 2 defines a selectivity for each value of f_n which reduces with increasing frequency. It also notes that the high value of the fundamental compared to the harmonic currents, or the possible high peak value of current compared to its rms value, must not overload the input stages of the instrument. The draft revision to the standard defines the requirements for time domain instrumentation rather differently to those for the frequency domain type.

When the harmonic components fluctuate while the measurement is being made, the response at the indicating output should be that of a first order low pass filter with a time constant of 1.5 seconds.

3.5.2 Test conditions

In general, the EUT should be operated by setting its user controls or program mode to give the maximum harmonic amplitude for each successive harmonic component in turn. Independent lamp dimmers and other phase-control devices should be set for a firing angle of 90°. For equipment with several separately controlled, independent circuits each circuit should be tested separately. For personal computers, auxiliary equipment which may be part of the overall system but is not permanently connected can be tested separately.

Part 2

Design Principles

Chapter 4

Interference coupling mechanisms

4.1 Source and victim

Situations in which the question of electromagnetic compatibility arises invariably have two complementary aspects. Any such situation must have a source of interference emissions and a victim which is susceptible to this interference. If either of these is not present, there is no EMC problem. If both source and victim are within the same piece of equipment we have an "intrasystem" EMC situation; if they are two different items, such as a computer monitor and a radio receiver, it is said to be an "inter-system" situation. The standards which were discussed in chapter 2 were all related to controlling inter-system EMC. The same equipment may be a source in one situation and a victim in another.

Knowledge of how the source emissions are coupled to the victim is essential, since a reduction in the coupling factor is often the only way to reduce interference effects, if a product is to continue to meet its performance specification. The two aspects are frequently reciprocal, that is measures taken to improve emissions will also improve the susceptibility, though this is not invariably so. For analysis, they are more easily considered separately.

Systems EMC

Putting source and victim together shows the potential interference routes that exist from one to the other (Figure 4.1). When systems are being built, you need to know the emissions signature and susceptibility of the component equipment, to determine whether problems are likely to be experienced with close coupling. Adherence to published emission and susceptibility standards does not guarantee freedom from systems EMC problems. Standards are written from the point of view of protecting a particular service – in the case of emissions standards, this is radio broadcast and telecommunications – and they have to assume a minimum separation between source and victim.

Most electronic hardware contains elements which are capable of antenna-like behaviour, such as cables, pcb tracks, internal wiring and mechanical structures. These elements can unintentionally transfer energy via electric, magnetic or electromagnetic fields which couple the circuits. In practical situations, intra-system and external coupling between equipment is modified by the presence of screening and dielectric materials, and by the layout and proximity of interfering and victim equipment and especially their respective cables. Ground or screening planes will enhance an interfering signal by reflection or attenuate it by absorption. Cable-to-cable coupling can be either capacitive or inductive and depends on orientation, length and proximity. Dielectric materials may also reduce the field by absorption, though this is negligible compared with the effects of conductors in most practical situations.

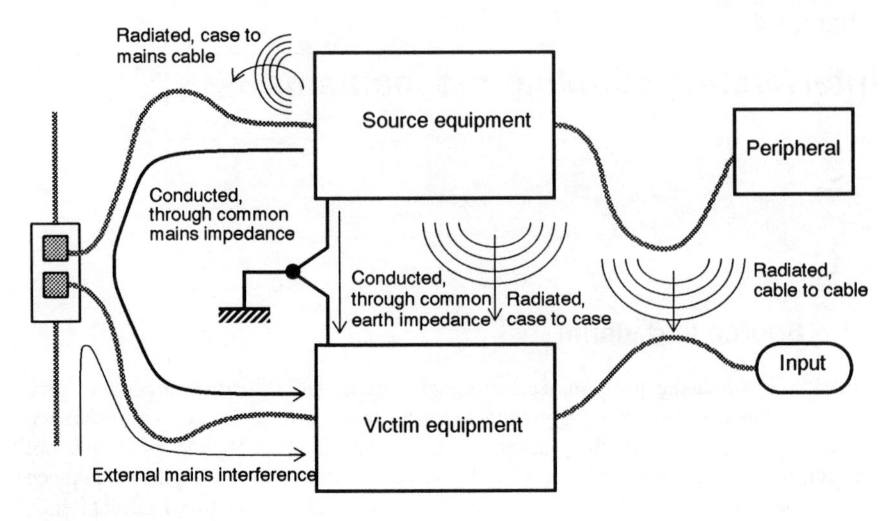

Figure 4.1 Coupling paths

4.1.1 Common impedance coupling

Common impedance coupling routes are those which are due to a circuit impedance which the source shares with the victim. The most obvious common impedances are those in which the impedance is physically present, as with a shared conductor; but the common impedance may also be due to mutual inductive coupling between two current loops, or to mutual capacitive coupling between two voltage nodes. Philosophically speaking, every node and every loop is coupled to all others throughout the universe. Practically, the strength of coupling falls off very rapidly with distance. Figure 4.4 shows the variation of mutual capacitance and inductance of a pair of parallel wires versus their separation, and the field equations in Appendix C (section C.3) give the precise expressions for the field at any point due to a radiating element.

4.1.1.1 Conductive connection

When an interference source (output of system A in Figure 4.2) shares a ground connection with a victim (input of system B) then any current due to A's output flowing through the common impedance section X-X develops a voltage in series with B's input. The common impedance need be no more than a length of wire or pcb track. High frequency or high di/dt components in the output will couple more efficiently because of the inductive nature of the impedance (see appendix C section C.5 for the inductance of various conductor configurations). The output and input may be part of the same system, in which case there is a spurious feedback path through the common impedance which can cause oscillation.

The solution as shown in Figure 4.2 is to separate the connections so that there is no common current path, and hence no common impedance, between the two circuits. The only "penalty" for doing this is the need for extra wiring or track to define the separate circuits. This applies to any circuit which may include a common impedance, such as power rail connections. Grounds are the most usual source of common impedance because the ground connection, often not shown on circuit diagrams, is taken for granted.

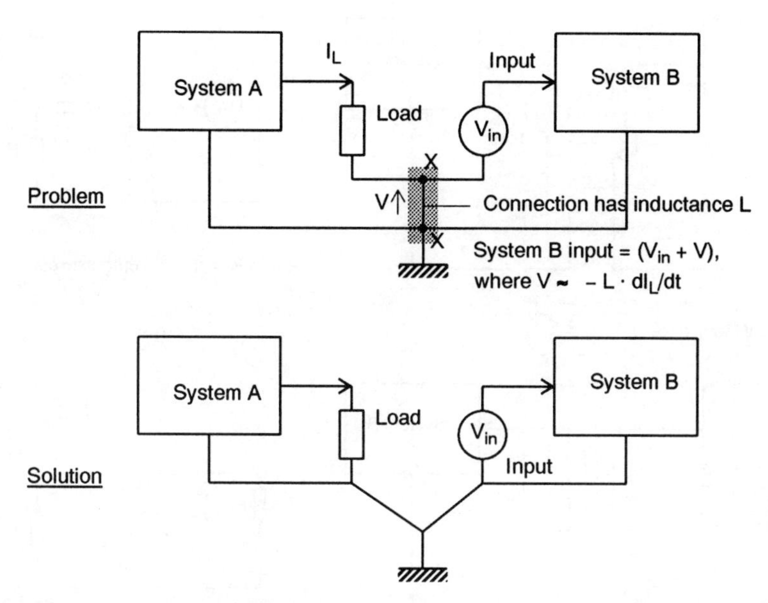

Figure 4.2 Conducted common impedance coupling

4.1.1.2 *Magnetic induction*

Alternating current flowing in a conductor creates a magnetic field which will couple with a nearby conductor and induce a voltage in it (Figure 4.3(a)). The voltage induced in the victim conductor is given by equation (4.1):

$$V = -M \cdot dI_L/dt \quad (4.1)$$

where M is the mutual inductance in henries

M depends on the areas of the source and victim current loops, their orientation and separation distance, and the presence of any magnetic screening. Appendix C (section C.5)gives mutual inductance formulae, but typical values for short lengths of cable loomed together lie in the range 0.1 to 3μH. The equivalent circuit for magnetic coupling is a voltage generator in series with the victim circuit. Note that the coupling is unaffected by the presence or absence of a direct connection between the two circuits; the induced voltage would be the same if both circuits were isolated or connected to ground.

4.1.1.3 *Electric induction*

Changing voltage on one conductor creates an electric field which may couple with a nearby conductor and induce a voltage on it (Figure 4.3(b)). The voltage induced on the victim conductor in this manner is

$$V = C_C \cdot dV_L/dt \cdot Z_{in}, \quad (4.2)$$

where C_C is the coupling capacitance and
Z_{in} is the impedance to ground of the victim circuit

This assumes that the impedance of the coupling capacitance is much higher than that of the circuit impedances. The noise is injected as if from a current source with a value

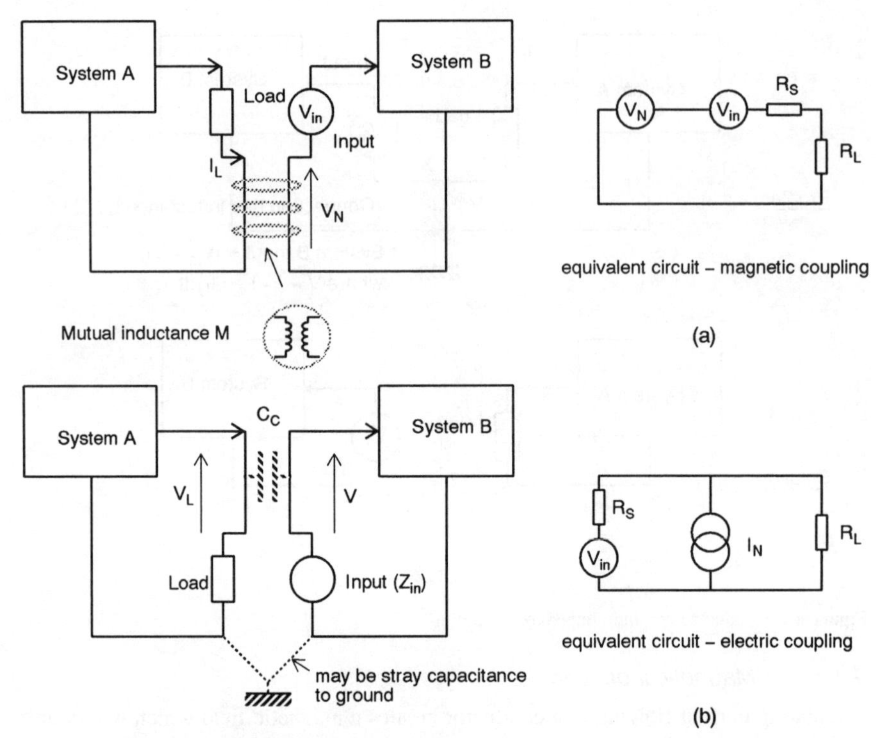

Figure 4.3 Magnetic and electric induction

of $C_C \cdot dV_L/dt$. The value of C_C is a function of the distance between the conductors, their effective areas and the presence of any electric screening material. Typically, two parallel insulated wires 0.1" apart show a coupling capacitance of about 50pF per metre; the primary-to-secondary capacitance of an unscreened medium power mains transformer is 100–1000pF.

Floating circuits

In this case, both circuits need to be referenced to ground for the coupling path to be complete. But if either is floating, this does *not* mean that there is no coupling path: the floating circuit will exhibit a stray capacitance to ground and this is in series with the direct coupling capacitance. Alternatively, there will be stray capacitance direct from the low-voltage nodes of A to B even in the absence of any ground node. The noise current will still be injected across R_L but its value will be determined by the series combination of C_C and the stray capacitance.

4.1.1.4 Effect of load resistance

Note that the difference in equivalent circuits for magnetic and electric coupling means that their behaviour with a varying circuit load resistance is different. Electric field coupling *increases* with an increasing R_L while magnetic field coupling *decreases* with an increasing R_L. This property can be useful for diagnostic purposes; if you vary R_L while observing the coupled voltage, you can deduce which mode of coupling predominates. For the same reason, magnetic coupling is more of a problem for low-impedance circuits while electric coupling applies to high impedance circuits.

4.1.1.5 Spacing

Both mutual capacitance and mutual inductance are affected by the physical separation of source and victim conductors. Figure 4.4 shows the effect of spacing on mutual capacitance of two parallel wires in free space, and on mutual inductance of two conductors over a ground plane (the ground plane provides a return path for the current in each circuit).

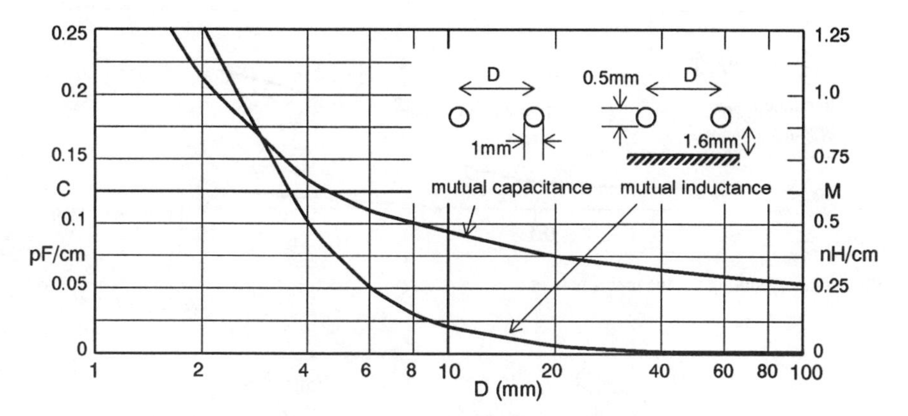

Figure 4.4 Mutual capacitance and inductance versus spacing

4.1.2 Mains coupling

Interference can propagate from a source to a victim via the mains distribution network to which both are connected. This is not well characterized at high frequencies, although the impedance viewed at any connection is reasonably predictable. We have already seen that the RF impedance presented by the mains can be approximated by a network of 50Ω in parallel with 50μH (section 3.1.2.2). For short distances such as via adjacent outlets on the same ring, coupling via the mains connection of two items of equipment can be represented by the equivalent circuit of Figure 4.5.

Over longer distances, power cables are fairly low loss transmission lines of around 150–200Ω characteristic impedance up to about 10MHz. However, in any local power distribution system the disturbances and discontinuities introduced by load connections, cable junctions and distribution components will dominate the RF transmission characteristic. These all tend to increase the attenuation.

4.1.3 Radiated coupling

To understand how energy is coupled from a source to a victim at a distance with no intervening connecting path, you need to have a basic understanding of electromagnetic wave propagation. This section will do no more than introduce the necessary concepts. The theory of EM waves has been well covered in many other works [3][6][11].

4.1.3.1 Field generation

An electric field (E field) is generated between two conductors at different potentials. The field is measured in volts per metre and is proportional to the applied voltage divided by the distance between the conductors.

A magnetic field (H field) is generated around a conductor carrying a current, is

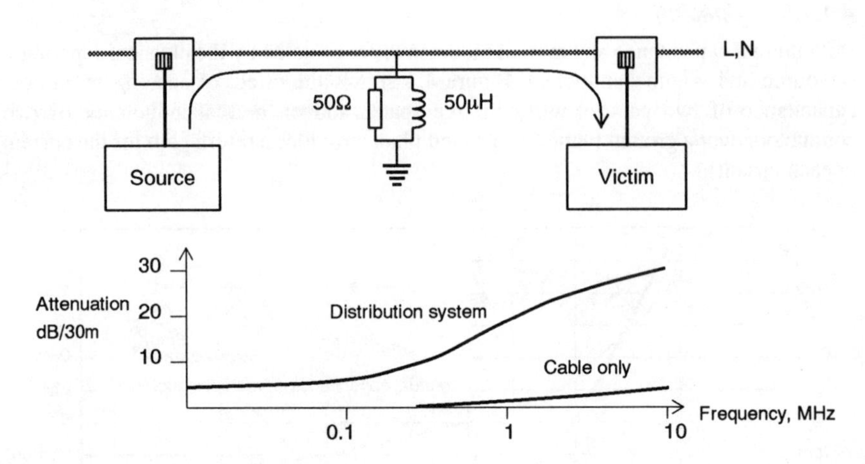

Figure 4.5 Coupling via the mains network

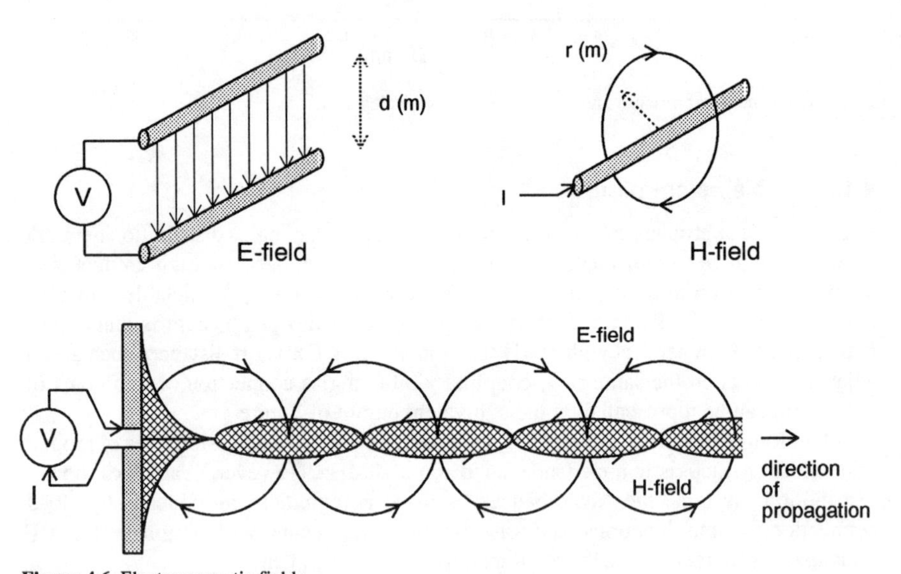

Figure 4.6 Electromagnetic fields

measured in amps per metre and is proportional to the current divided by the distance from the conductor.

When an alternating voltage generates an alternating current through a network of conductors, an electromagnetic (EM) wave is generated which propagates as a combination of E and H fields at right angles. The speed of propagation is determined by the medium; in free space it is equal to the speed of light, 3.10^8m/s. Near to the radiating source the geometry and strength of the fields depend on the characteristics of the source. Further away only the orthogonal fields remain. Figure 4.6 demonstrates these concepts graphically. The field equations are presented in Appendix C (section C.3).

4.1.3.2 Wave impedance

The ratio of the electric to magnetic field strengths (E/H) is called the wave impedance (Figure 4.7). The wave impedance is a key parameter of any given wave as it determines the efficiency of coupling with another conducting structure, and also the effectiveness of any conducting screen which is used to block it. In the far field, $d > \lambda/2\pi$, the wave is known as a plane wave and its impedance is constant and equal to the impedance of free space given by equation (4.3):

$$Z_o = (\mu_o/\varepsilon_o)^{0.5} = 120\pi = 377\Omega \qquad (4.3)$$

where μ_o is $4\pi \cdot 10^{-7}$ H/m
and ε_o is $8.85 \cdot 10^{-12}$ F/m

In the near field, $d < \lambda/2\pi$, the wave impedance is determined by the characteristics of the source. A low current, high voltage radiator (such as a rod) will generate mainly an electric field of high impedance, while a high current, low voltage radiator (such as a loop) will generate mainly a magnetic field of low impedance. If the radiating structure happens to have an impedance around 377Ω, then a plane wave can in fact be generated in the near field, depending on geometry.

The region around $\lambda/2\pi$, or approximately one sixth of a wavelength, is the transition region between near and far fields. Plane waves are always assumed to be in the far field, while if individual electric or magnetic fields are being considered they are assumed to be in the near field. Appendix C presents the formulae that underpin this description.

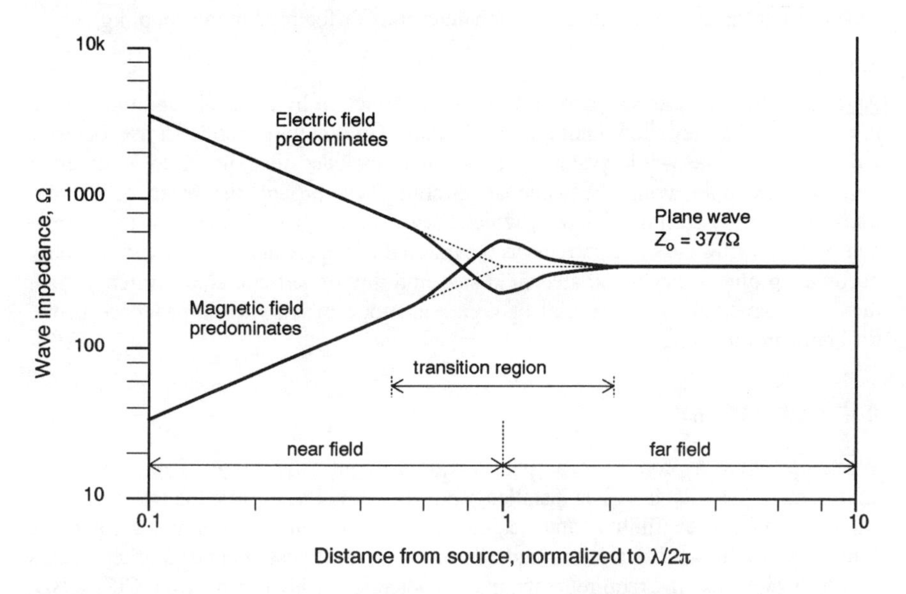

Figure 4.7 The wave impedance

4.1.3.3 Coupling modes

The concepts of differential mode, common mode and antenna mode radiated field coupling are fundamental to an understanding of EMC and will crop up in a variety of guises throughout this book. They apply to coupling of both emissions and incoming interference.

Differential mode

Consider two items of equipment interconnected by a cable (Figure 4.8). The cable carries signal currents in differential mode (go and return) down the two wires in close proximity. A radiated field can couple to this system and induce differential mode interference between the two wires; similarly, the differential current will induce a radiated field of its own. The ground reference plane (which may be external to the equipment or may be formed by its supporting structure) plays no part in the coupling.

Common mode

The cable also carries currents in common mode, that is, all flowing in the same direction on each wire. These currents very often ***have nothing at all to do with the signal currents***. They may be induced by an external field coupling to the loop formed by the cable, the ground plane and the various impedances connecting the equipment to ground, and may then result in internal differential currents to which the equipment is susceptible. Alternatively, they may be generated by internal noise voltages between the ground reference point and the cable connection, and be responsible for radiated emissions. Notice that the stray capacitances and inductances associated with the wiring and enclosure of each unit are an integral part of the common mode coupling circuit, and play a large part in determining the amplitude and spectral distribution of the common mode currents. These stray reactances are incidental rather than designed in to the equipment and are therefore much harder to control or predict than parameters such as cable spacing and filtering which determine differential mode coupling.

Antenna mode

Antenna mode currents are carried in the same direction by the cable and the ground reference plane. They should not arise as a result of internally-generated noise, but they will flow when the whole system, ground plane included, is exposed to an external field. An example would be when an aircraft flies through the beam of a radar transmission; the aircraft structure, which serves as the ground plane for its internal equipment, carries the same currents as the internal wiring. Antenna mode currents only become a problem for the radiated field susceptibility of self-contained systems when they are converted to differential or common mode by varying impedances in the different current paths.

4.2 Emissions

When designing a product to a specification without knowledge of the system or environment in which it will be installed, you will normally separate the two aspects of emissions and susceptibility, and design to meet minimum requirements for each. Limits are laid down in various standards but individual customers or market sectors may have more specific requirements. In those standards which derive from CISPR (see chapter 2), emissions are subdivided into radiated emissions from the system as a whole, and conducted emissions present as common-mode currents on the interface and power cables. Conventionally, the breakpoint between radiated (high frequency) and conducted (low frequency) is set at 30MHz. Radiated emissions can themselves be separated into emissions that derive from internal pcbs or other wiring, and emissions from common-mode currents that find their way onto external cables that are connected to the equipment.

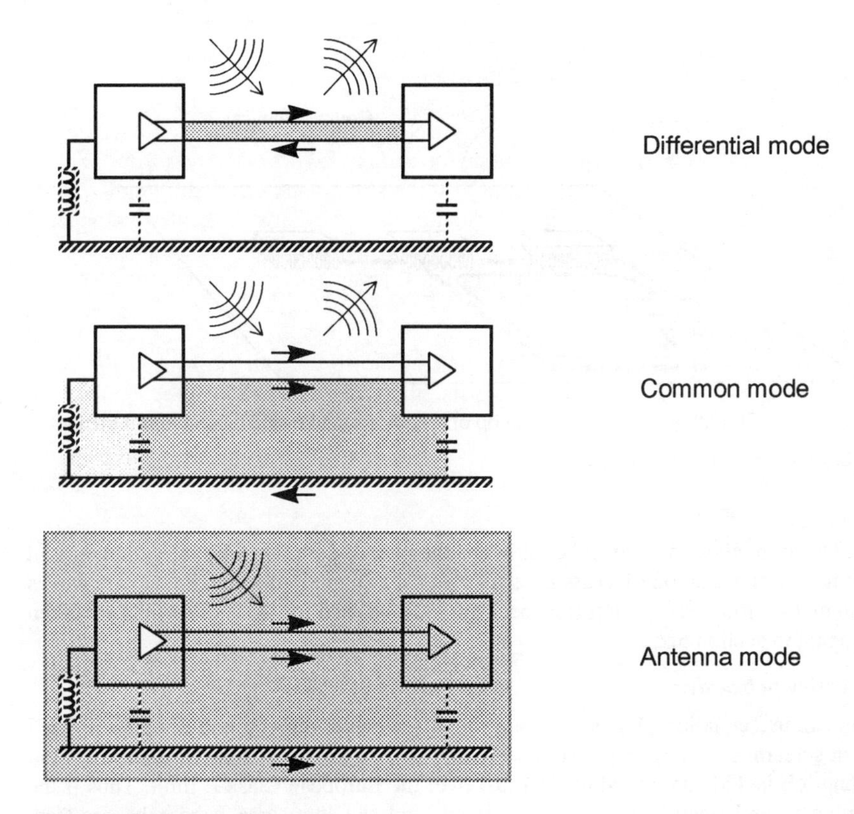

Figure 4.8 Radiated coupling modes

4.2.1 Radiated emissions

4.2.1.1 Radiation from the pcb

In most equipment, the primary emission sources are currents flowing in circuits (clocks, video and data drivers, and other oscillators) that are mounted on printed circuit boards. Radiated emission from a pcb can be modelled as a small loop antenna carrying the interference current (Figure 4.9). A small loop is one whose dimensions are smaller than a quarter wavelength (λ/4) at the frequency of interest (e.g. 1 metre at 75MHz). Most pcb loops count as "small" at emission frequencies of up to a few hundred MHz. When the dimensions approach λ/4 the currents at different points on the loop appear out of phase at a distance, so that the effect is to reduce the field strength at any given point. The maximum electric field strength from such a loop over a ground plane at 10 metres distance is proportional to the square of the frequency:

$$E = 263 \cdot 10^{-12} (f^2 \cdot A \cdot I_S) \text{ volts per metre [10]} \quad (4.4)$$

where A is the loop area in cm^2,
f (MHz) is the frequency of I_S the source current in mA.

In free space, the field falls off proportionally to distance from the source. The figure of 10m is used as this is the standard measurement distance for the European radiated emissions standards. A factor of 2 times is allowed for worst case field reinforcement due to reflection from the ground plane, which is also a required feature of testing to

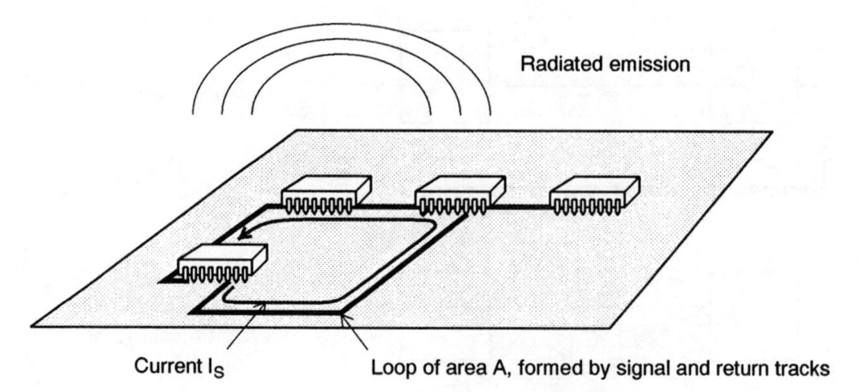

Figure 4.9 PCB radiated emissions

standards.

The loop whose area must be known is the overall path taken by the signal current and its return. Equation (4.4) assumes that I_S is at a single frequency. For square waves with many harmonics, the Fourier spectrum must be used for I_S. These points are taken up again in section 5.2.2.

Assessing pcb design

You can use equation (4.4) to indicate roughly whether a given pcb design will need extra screening. For example if A = 10cm^2, I_S = 20mA, f = 50MHz then the field strength E is 42dBμV/m, which is 12dB over the European Class B limit. Thus if the frequency and operating current are fixed, and the loop area cannot be reduced, screening will be necessary.

But the converse is not true. Differential mode radiation from small loops on pcbs is by no means the only contributor to radiated emissions; common mode currents flowing on the pcb and more importantly on attached cables can contribute much more. Paul [70] goes so far as to say

> "predictions of radiated emissions based solely on differential-mode currents will generally bear no resemblance to measured levels of radiated emissions. Therefore, basing system EMC design on differential-mode currents and the associated prediction models that use them exclusively while neglecting to consider the (usually much larger) emissions due to common-mode currents can lead to a strong 'false sense of security'".

Common mode currents on the pcb itself are not at all easy to predict, in contrast with the differential mode currents which are governed by Kirchoff's current law. The return path for common mode currents is via stray capacitance (displacement current) to other nearby objects and therefore a full prediction would have to take the detailed mechanical structure of the pcb and its case, as well as its proximity to ground and to other equipment, into account. Except for trivial cases this is to all intents and purposes impossible. It is for this reason more than any other that EMC design has earned itself the distinction of being a "black art".

4.2.1.2 Radiation from cables

Fortunately (from some viewpoints) radiated coupling at VHF tends to be dominated by cable emissions, rather than by direct radiation from the pcb. This is for the simple reason that typical cables resonate in the 30 – 100MHz region and their radiating efficiency is higher than pcb structures at these frequencies. The interference current is generated in common mode from ground noise developed across the pcb or elsewhere in the equipment and may flow along the conductors, or along the shield of a shielded cable. The model for cable radiation at lower frequencies (Figure 4.10) is a short

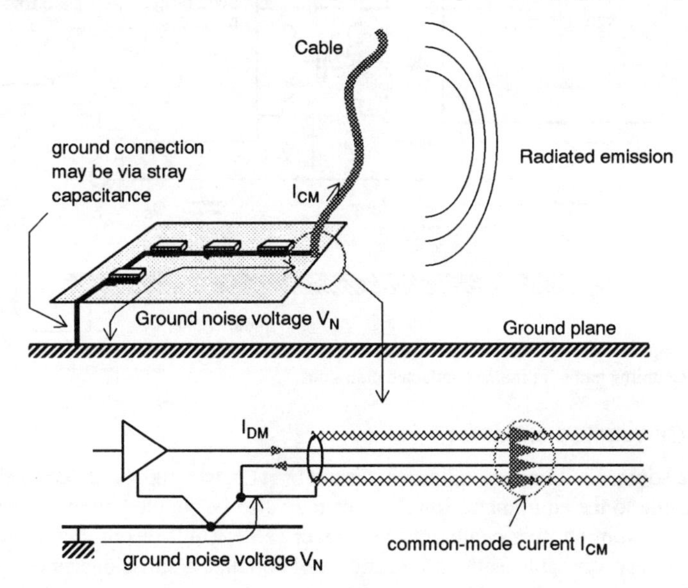

Figure 4.10 Cable radiated emissions

($L < \lambda/4$) monopole antenna over a ground plane. (When the cable length is resonant the model becomes invalid.) The maximum field strength allowing +6dB for ground plane reflections at 10m due to this radiation is directly proportional to frequency:

$$E = 1.26 \cdot 10^{-4} \cdot (f \cdot L \cdot I_{CM}) \text{ volts per metre [10]} \quad (4.5)$$

where L is the cable length in metres and
I_{CM} is the common-mode current at f MHz in mA flowing in the cable.

For a 1m cable, I_{CM} must be less than 20μA for a field strength at 10m of 42dBμV/m – i.e., a thousand times less than the equivalent differential mode current!

4.2.1.3 Common mode cable noise

At the risk of repetition, it is vital to appreciate the difference between common mode and differential mode cable currents. Differential mode current, I_{DM} in Figure 4.10, is the current which flows in one direction along one cable conductor and in the reverse direction along another. It is normally equal to the signal or power current, and is not present on the shield. It contributes little to the net radiation as long as the total loop area formed by the two conductors is small; the two currents tend to cancel each other. Common mode current I_{CM} flows equally in the same direction along all conductors in the cable, potentially including the shield, and is quite unrelated to the signal currents. It returns via the associated ground network and therefore the radiating loop area is large and uncontrolled. As a result, even a small I_{CM} can result in large emitted signals.

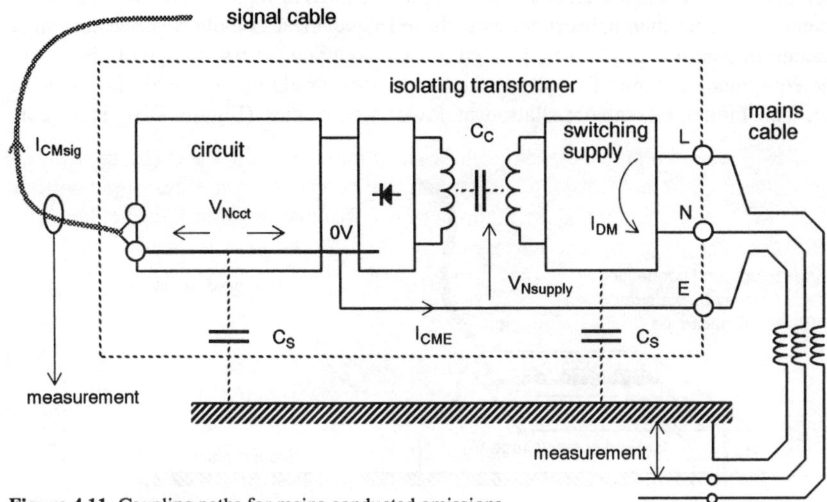

Figure 4.11 Coupling paths for mains conducted emissions

4.2.2 Conducted emissions

Interference sources within the equipment circuit or its power supply are coupled onto the power cable to the equipment. Interference may also be coupled either inductively or capacitively from another cable onto the power cable. Until recently, attention has focussed on the power cable as the prime source of conducted emissions since CISPR-based standards have only specified measurements on this cable. However, signal and control cables can and do also act as coupling paths and amendments to the standards will apply measurements to these cables as well.

The resulting interference may appear as differential mode (between live and neutral, or between signal wires) or as common mode (between live/neutral/signal and earth) or as a mixture of both. For signal and control lines, only common mode currents are regulated. For the mains port, the voltages between live and earth and between neutral and earth at the far end of the mains cable are measured. Differential mode emissions are normally associated with low-frequency switching noise from the power supply, while common-mode emissions can be due to the higher frequency switching components, internal circuit sources or inter-cable coupling.

4.2.2.1 Coupling paths

The equivalent circuit for a typical product with a switchmode power supply, shown in Figure 4.11, gives an idea of the various paths these emissions can take. (Section 5.2.4 looks at SMPS emissions in more detail.) Differential mode current I_{DM} generated at the input of the switching supply is converted by imbalances in stray capacitance, and by the mutual inductance of the conductors in the mains cable, into interference voltages with respect to earth at the measurement point. Higher frequency switching noise components $V_{Nsupply}$ are coupled through C_C to appear between L/N and E on the mains cable, and C_S to appear with respect to the ground plane. Circuit ground noise V_{Ncct} (digital noise and clock harmonics) is referenced to ground by C_S and coupled out via signal cables as I_{CMsig} or via the safety earth as I_{CME}.

The problem in a real situation is that all these mechanisms are operating simultaneously, and the stray capacitances C_S are widely distributed and unpredictable, depending heavily on proximity to other objects if the case is unscreened. A partially-screened enclosure may actually worsen the coupling because of its higher capacitance to the environment.

4.2.3 Mains harmonics

One EMC phenomenon, which comes under the umbrella of the EMC Directive and is usually classified as an "emission", is the harmonic content of the mains input current. This is mildly confusing since the equipment is not actually "emitting" anything: it is simply drawing its power at harmonics of the line frequency as well as at the fundamental.

4.2.3.1 The supplier's problem

The problem of mains harmonics is principally one for the supply authorities, who are mandated to provide a high quality electricity supply. If the aggregate load at a particular mains distribution point has a high harmonic content, the non-zero distribution source impedance will cause distortion of the voltage waveform at this point. This in turn may cause problems for other users connected to that point, and the currents themselves may also create problems (such as overheating of transformers and compensating components) for the supplier. The supplier does of course have the option of uprating the distribution components or installing special protection measures, but this is expensive and the supplier has room to argue that the users should bear some of the costs of the pollution they create.

Throughout the 1980s, harmonic pollution has been increasing and it has been principally due to low power electronic loads installed in large numbers. Between them, domestic TV sets and office information technology equipment account for about 80% of the problem. Other types of load which also take significant harmonic currents are not widely enough distributed to cause a serious problem yet, or are dealt with individually at the point of installation as in the case of industrial plant. The supply authorities are nevertheless sufficiently worried to want to extend harmonic emission limits to all classes of electronic products.

4.2.3.2 Non-linear loads

A plain resistive load across the mains draws current only at the fundamental frequency (50Hz in Europe). Most electronic circuits are anything but resistive. The universal rectifier-capacitor input draws a high current at the peak of the voltage waveform and zero current at other times; the well known triac phase control method for power control (lights, motors, heaters etc.) begins to draw current only partway through each half cycle. These current waveforms can be represented as a Fourier series, and it is the harmonic amplitudes of the series that are subject to regulation. The relevant standard is EN 60 555 part 2 (see section 2.2.8), which in its present (1987) version applies only to household products.

There is a proposal to extend the scope of EN 60 555 to cover a wide range of products, and it will affect virtually all mains powered electronic equipment above a certain power level which has a rectifier-reservoir input. The harmonic limits are effectively an additional design constraint on the values of the input components, most notably the input series impedance (which is not usually considered as a desirable input component at all). Figure 4.12(a), which is a Fourier analysis of the current waveform calculated in the time domain, shows the harmonic content of input current for a rectifier-reservoir combination with a fairly high series resistance. This value of series

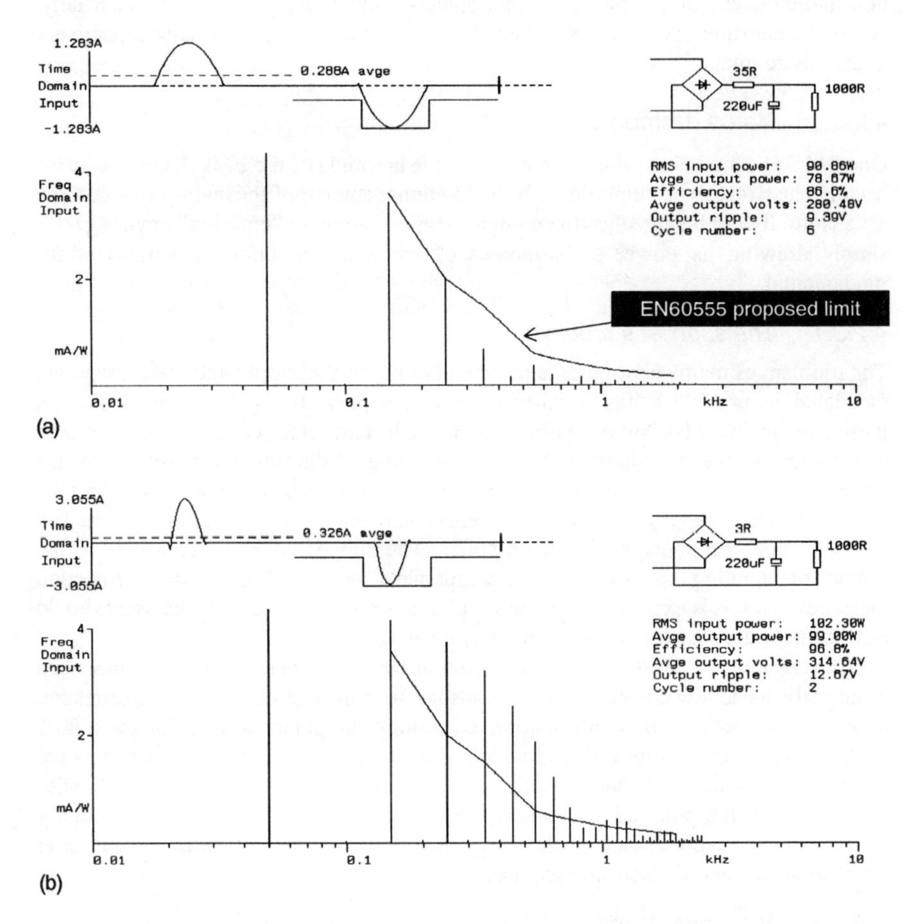

Figure 4.12 Mains input current harmonics for rectifier-reservoir circuit

resistance would not normally be found except with very inefficient transformer-input supplies. The fifth harmonic content just manages to meet the limit proposed in the draft revision of EN 60 555-2 (but note that these limits have not been accepted and are likely to change in the final revision).

4.2.3.3 The effect of series resistance

Figure 4.12(b) illustrates the difference in input harmonics resulting from a tenfold reduction in input resistance. This level of input resistance would be typical for a direct-off-line switching supply and many highly efficient supplies could boast a lower R_S. The peak input current has increased markedly while its duty cycle has shrunk, leading to a much higher crest factor (ratio of peak to root mean square current) and thus higher levels of harmonics.

Increasing input series resistance to meet the harmonic limits is expensive in terms of power dissipation except at very low powers. In practice, deliberately dissipating between 10 and 20% of the input power rapidly becomes unreasonable above levels of

50–100W. Alternatives are to include a series input choke, which since it must operate down to 50Hz is expensive in size and weight; or to include electronic power factor correction (PFC), which converts the current waveform to a near-sinusoid but is expensive in cost and complexity. PFC is essentially a switchmode converter on the front-end of the supply and therefore is likely to contribute extra RF switching noise at the same time as it reduces input current harmonics. It is possible to combine PFC with the other features of a direct-off-line switching supply, so that if you are intending to use a SMPS anyway there will be little extra penalty. It also fits well with other contemporary design requirements such as the need for a "universal" (90–260V) input voltage range. Such power supplies can already be bought off-the-shelf, but unless you are a power supply specialist, to design a PFC-SMPS yourself will take considerable extra design and development effort.

4.2.3.4 Phase control

Power control circuits which vary the switch-on point with the phase of the mains waveform are another major source of harmonic distortion on the input current. Lighting controllers are the leading example of these. Figure 4.13 shows the harmonic content of such a waveform switched at 90° (the peak of the cycle, corresponding to half power). The maximum harmonic content occurs at this point, decreasing as the phase is varied either side of 90°. Whether lighting dimmers will comply with the draft limits

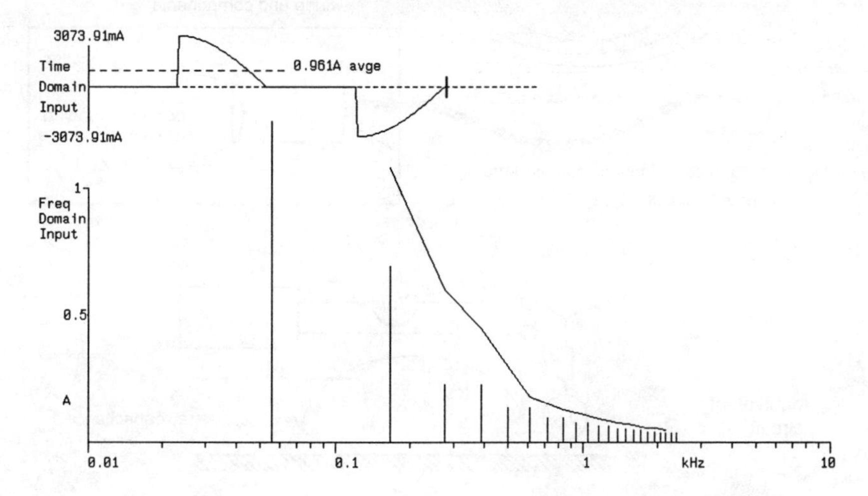

Figure 4.13 Mains input current harmonics for 500W phase control circuit at half power

in EN 60 555-2 without input filtering or PFC depends at present on their power level, since these limits are set at an absolute value. As the draft stands, unmodified dimmers of greater than about 5A rating would be outlawed.

4.3 Susceptibility

Electronic equipment will be susceptible to environmental electromagnetic fields and/ or to disturbances coupled into its ports via connected cables. An electrostatic discharge may be coupled in via the cables or the equipment case, or a nearby discharge can create a local field which couples directly with the equipment. The potential threats are

- radiated RF fields
- conducted transients
- electrostatic discharge (ESD)
- magnetic fields
- supply voltage disturbances

Quite apart from legal requirements, equipment that is designed to be immune to these effects – especially ESD and transients – will save its manufacturer considerable expense through preventing field returns. Unfortunately, the shielding and circuit suppression measures that are required for protection against ESD or RF interference may be more than you need for emission control.

4.3.1 Radiated field

An external field can couple either directly with the internal circuitry and wiring in differential mode or with the cables to induce a common mode current (Figure 4.14). Coupling with internal wiring and pcb tracks is most efficient at frequencies above a

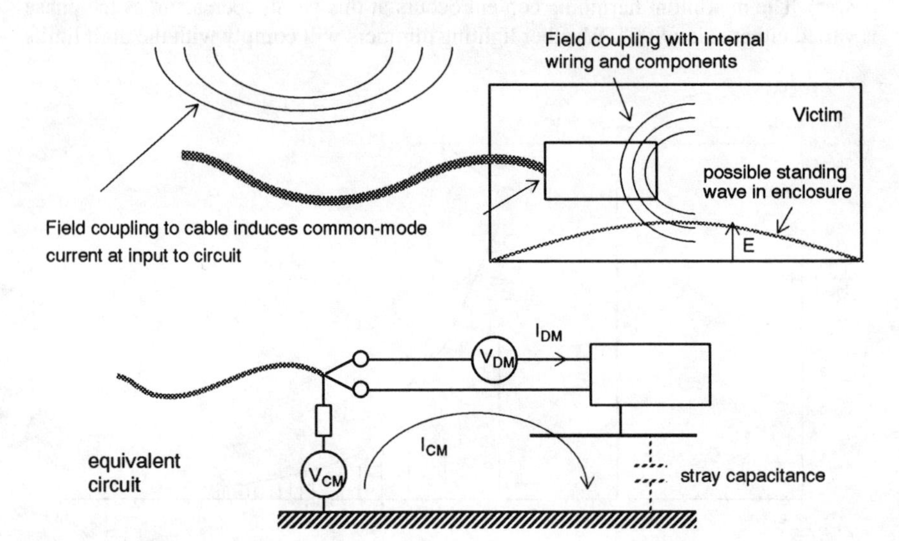

Figure 4.14 Radiated field coupling

few hundred MHz, since wiring lengths of a few inches approach resonance at these frequencies.

RF voltages or currents in analogue circuits can induce nonlinearity, overload or DC bias and in digital circuits can corrupt data transfer [73]. Modulated fields can have greater effect than unmodulated ones. Likely sources of radiated fields are walkie-talkies, cellphones, high-power broadcast transmitters and radars. Field strengths between 1 and 10V/m from 20MHz to 1GHz are typical, and higher field strengths can occur in environments close to such sources.

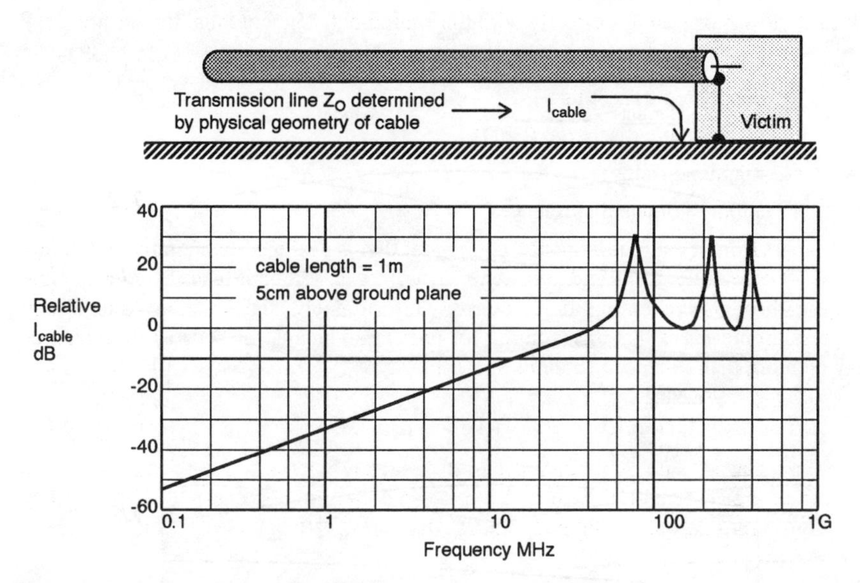

Figure 4.15 Cable coupling to radiated field

4.3.1.1 Cable resonance

Cables are most efficient at coupling RF energy into equipment at the lower end of the vhf spectrum (30 – 100MHz). The external field induces a common mode current on the cable shield or on all the cable conductors together, if it is unshielded. The common mode current effects in typical installations tend to dominate the direct field interactions with the equipment as long as the equipment's dimensions are small compared with half the wavelength of the interfering signal.

A cable connected to a grounded victim equipment can be modelled as a single conductor over a ground plane, which appears as a transmission line (Figure 4.15). The current induced in such a transmission line by an external field increases steadily with frequency until the first resonance is reached, after which it exhibits a series of peaks and nulls at higher resonances [15]. The coupling mechanism is enhanced at the resonant frequency of the cable, which depends on its length and on the reactive loading of whatever equipment is attached to its end. A length of 2 metres is quarter-wave resonant at 37.5MHz, half-wave resonant at 75MHz.

Cable loading

The dominant resonant mode depends on the RF impedance (high or low) at the distant end of the cable. If the cable is connected to an ungrounded object such as a hand controller it will have a high RF impedance, which will cause a high coupled current at quarter-wave resonance and high coupled voltage at half-wave. Extra capacitive loading such as body capacitance will lower its apparent resonant frequency.

Conversely, a cable connected to another grounded object such as a separately earthed peripheral will see a low impedance at the far end, which will generate high coupled current at half-wave and high coupled voltage at quarter-wave resonance. Extra inductive loading, such as the inductance of the earth connection, will again tend to lower the resonant frequency.

These effects are summarized in Figure 4.16. The RF common mode impedance of

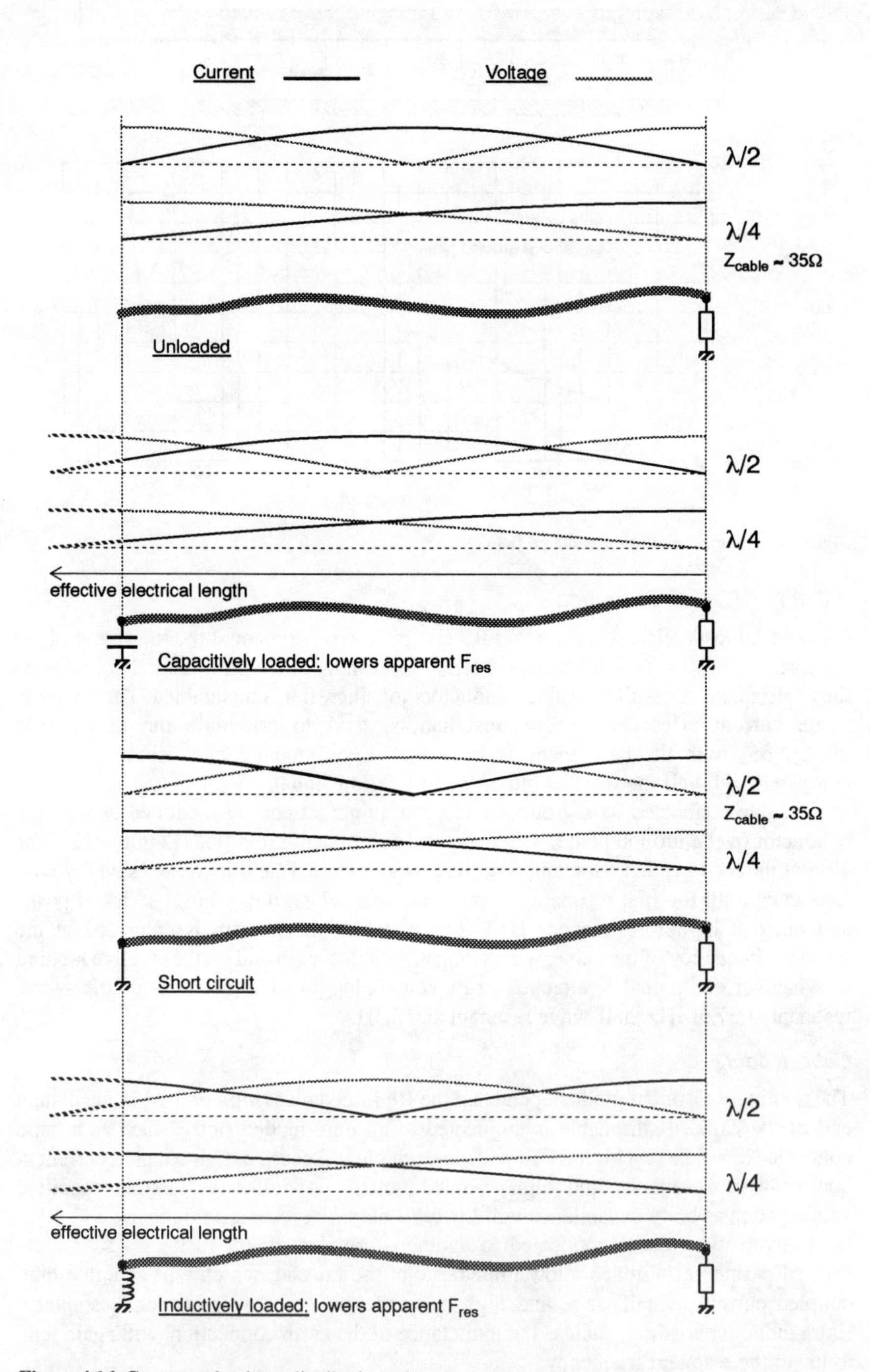

Figure 4.16 Current and voltage distributions along a resonant cable

the cable varies from around 35Ω at quarter wave resonance to several hundred ohms maximum. A convenient average figure (and one that is taken in many standards) is 150Ω. Because cable configuration, layout and proximity to grounded objects are outside the designer's control, attempts to predict resonances and impedances accurately are unrewarding.

4.3.1.2 Current injection

A convenient method for testing the RF susceptibility of equipment without reference to its cable configuration is to inject RF as a common mode current or voltage directly onto the cable port (see also section 3.2.4)[56]. This represents real-life coupling situations at lower frequencies well, until the equipment dimensions approach a half wavelength. It can also reproduce the fields (E_{RF} and H_{RF}) associated with radiated field coupling. The route taken by the interference currents, and hence their effect on the circuitry, depends on the various internal and external RF impedances to earth, as shown in Figure 4.17. Connecting other cables will modify the current flow to a marked extent, especially if the extra cables interface to a physically different location on the pcb or equipment. An applied voltage of 1V, or an injected current of 10mA, corresponds in typical cases to a radiated field strength of 1V/m.

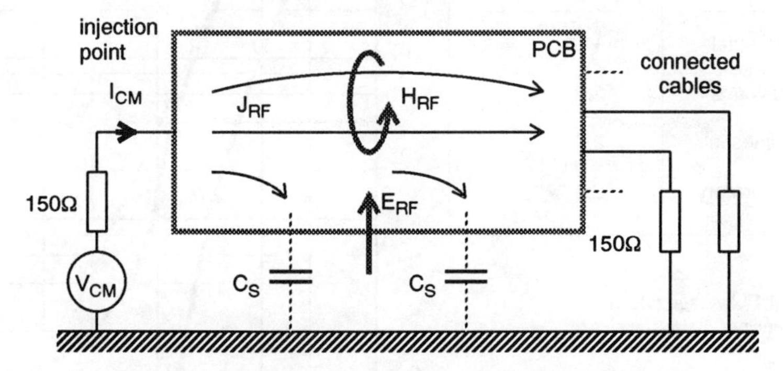

Figure 4.17 Common mode RF injection

4.3.1.3 Cavity resonance

A screened enclosure can form a resonant cavity; standing waves in the field form between opposite sides when the dimension between the sides is a multiple of a half-wavelength. The electric field is enhanced in the middle of this cavity while the magnetic field is enhanced at the sides. This effect is usually responsible for peaks in the susceptibility versus frequency profile in the UHF region.

Predicting the resonant frequency accurately from the enclosure dimensions is rarely successful because the contents of the enclosure tend to "detune" the resonance. But for an empty cavity, resonances occur at

$$F = 150 \cdot \sqrt{(k/l)^2 + (m/h)^2 + (n/w)^2} \text{ MHz} \quad [17] \qquad (4.6)$$

where l, h and w are the enclosure dimensions in metres
k, m and n are positive integers, but no more than one at a time can be zero

For approximately equal enclosure dimensions the lowest possible resonant frequency will be given by equation (4.7):

$$F \approx 212/l \approx 212/h \approx 212/w \text{ MHz} \tag{4.7}$$

4.3.2 Transients

Transient overvoltages occur on the mains supply leads due to switching operations, fault clearance or lightning strikes elsewhere on the network. Transients over 1kV account for about 0.1% of the total number of transients observed. A study by the German ZVEI [40] made a statistical survey of 28,000 live-to-earth transients exceeding 100V, at 40 locations over a total measuring time of about 3,400 hours. Their results were analyzed for peak amplitude, rate of rise and energy content. Table 4.1 shows the average rate of occurrence of transients for four classes of location, and Figure 4.18 shows the relative number of transients as a function of maximum transient amplitude. This shows that the number of transients varies roughly in inverse proportion to the cube of peak voltage.

Area Class	Average rate of occurrence (transients/hour)
Industrial	17.5
Business	2.8
Domestic	0.6
Laboratory	2.3

Table 4.1 Average rate of occurrence of mains transients

Sources:
Transients in Low Voltage Supply Networks, J.J.Goedbloed, IEEE Transactions on Electromagnetic Compatibility, Vol EMC-29 No 2, May 1987, p 107
Characterization of Transient and CW Disturbances Induced in Telephone Subscriber Lines, J.J.Goedbloed, W.A. Pasmooij, IEE 7th International Conference on EMC, York 1990

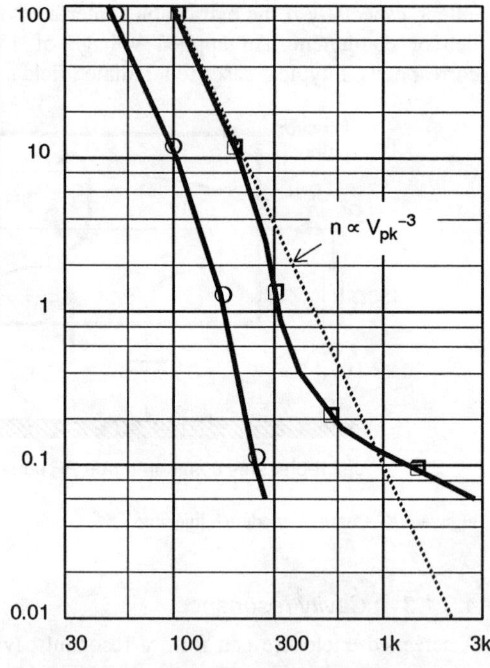

Figure 4.18 Relative number of transients (percent) vs. maximum transient amplitude (volts)
❐ Mains lines (V_T 100V), ○ Telecomm lines (V_T 50V)

High energy transients may threaten active devices in the equipment power supply. Fast-rising edges are the most disruptive to circuit operation, since they are attenuated least by the coupling paths and they can generate large voltages in inductive ground and signal paths. The ZVEI study found that rate of rise increased roughly in proportion to the square root of peak voltage, being typically 3V/ns for 200V pulses and 10V/ns for 2kV pulses. Other field experience has shown that mechanical switching produces multiple transients (bursts) with risetimes as short as a few nanoseconds and peak amplitudes of several hundred volts. Attenuation through the mains network (see section 4.1.2) restricts fast risetime pulses to those generated locally.

Analogue circuits are almost immune to isolated short transients, whereas digital circuits are easily corrupted by them. As a general guide, microprocessor equipment should be tested to withstand pulses at least up to 2kV peak amplitude. Thresholds below 1kV will give unacceptably frequent corruptions in nearly all environments, while between 1kV - 2kV occasional corruption will occur. For a belt-and-braces approach for high reliability equipment, a 4 – 6kV threshold is recommended.

4.3.2.1 Coupling mode

Mains transients may appear in differential mode (symmetrically between live and neutral) or common mode (asymmetrically between live/neutral and earth). Coupling betwen the conductors in a supply network tends to mix the two modes. Differential mode spikes are usually associated with relatively slow risetimes and high energy, and require suppression to prevent input circuit damage but do not, provided this suppression is incorporated, affect circuit operation significantly. Common mode transients are harder to suppress because they require connection of suppression components between live and earth, or in series with the earth lead, and because stray capacitances to earth are harder to control. Their coupling paths are very similar to those followed by common mode RF signals. Unfortunately, they are also more damaging because they result in transient current flow in ground traces.

4.3.2.2 Transients on signal lines

Fast transients can be coupled, usually capacitively, onto signal cables in common mode, especially if the cable passes close to or is routed alongside an impulsive interference source. Although such transients are generally lower in amplitude than mains-borne ones, they are coupled directly into the I/O ports of the circuit and will therefore flow in the circuit ground traces.

Other sources of conducted transients are telecommunication lines and the automotive 12V supply. The automotive environment can regularly experience transients that are many times the nominal supply range. The most serious automotive transients (Figure 4.19) are the load dump, which occurs when the alternator load is suddenly disconnected during heavy charging; switching of inductive loads, such as motors and solenoids; and alternator field decay, which generates a negative voltage spike when the ignition switch is turned off. A recent standard (ISO 7637) has been issued to specify transient testing in the automotive field.

Work on common mode transients on telephone subscriber lines [41] has shown that the amplitude versus rate of occurrence distribution also follows a roughly inverse cubic law as in Figure 4.18. Actual amplitudes were lower than those on the mains (peak amplitudes rarely exceeded 300V). A transient ringing frequency of 1MHz and rise times of 10–20ns were found to be typical.

4.3.3 Electrostatic discharge

When two non-conductive materials are rubbed together, electrons from one material are transferred to the other. This results in the accumulation of triboelectric charge on the surface of the material. The amount of the charge caused by movement of the materials is a function of the separation of the materials in the triboelectric series (Figure 4.20(a)). Additional factors are the closeness of contact, rate of separation and

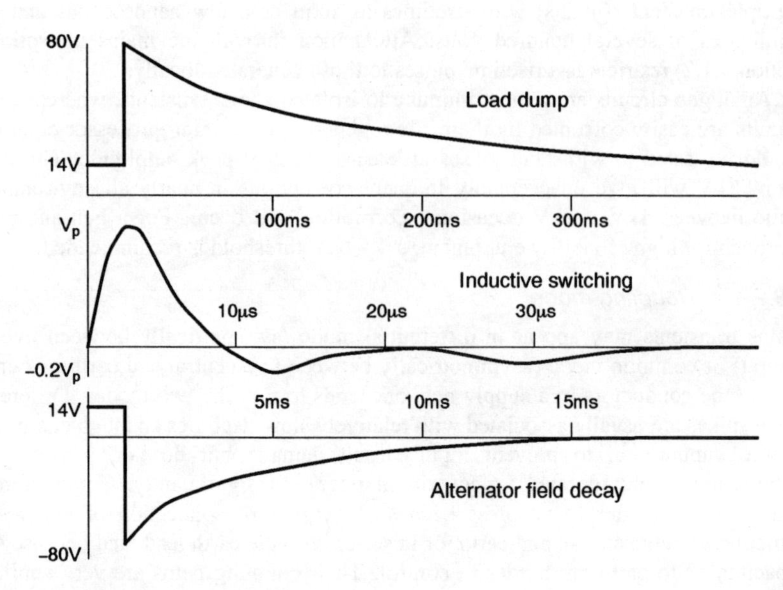

Figure 4.19 Automotive transients

humidity. The human body can be charged by triboelectric induction to several kV.

When the body (in the worst case, holding a metal object such as a key) approaches a conductive object, the charge is transferred to that object normally via a spark, when the potential gradient across the narrowing air gap is high enough. The energy involved in the charge transfer may be low enough to be imperceptible to the subject; at the other extreme it can be extremely painful.

4.3.3.1 The ESD waveform

When an electrostatically charged object is brought close to a grounded target the resultant discharge current consists of a very fast (sub-nanosecond) edge followed by a comparatively slow bulk discharge curve. The characteristic of the hand/metal ESD current waveform is a function of the approach speed, the voltage, the geometry of the electrode and the relative humidity. The equivalent circuit for such a situation is shown in Figure 4.20(c). The capacitance C_D is charged via a high resistance up to the electrostatic voltage V. The actual value of V will vary as the charging and leakage paths change with the environmental circumstances and movements of the subject. When a discharge is initiated, the free space capacitance C_S, which is directly across the discharge point, produces an initial current peak the value of which is only limited by the circuit inductance, while the main discharge current is limited by the discharge resistance R_D.

The resultant sub-nanosecond transient equalizing current of several tens of amps follows a complex route to ground through the equipment and is very likely to upset digital circuit operation if it passes through the circuit tracks. The paths are defined more by stray capacitance, case bonding and track or wiring inductance than by the designer's intended circuit. The high magnetic field associated with the current can induce transient voltages in nearby conductors that are not actually in the path of the current. Even if not discharged directly to the equipment, a nearby discharge such as to a metal desk or chair will generate an intense radiated field which will couple into

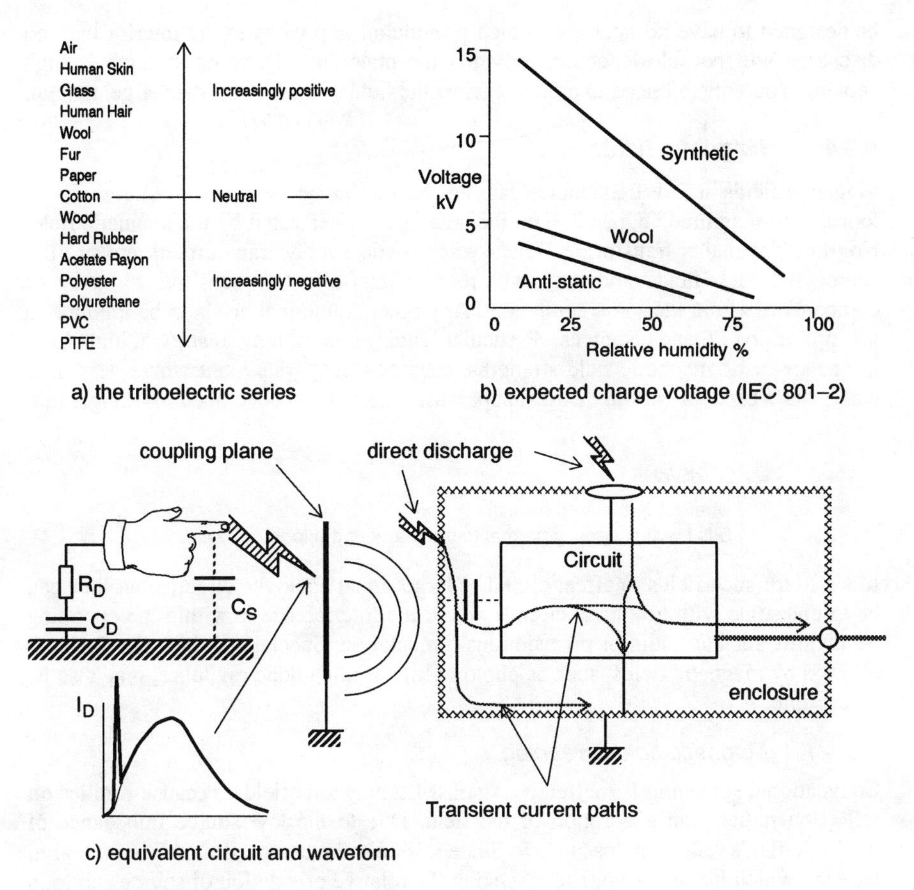

Figure 4.20 The electrostatic discharge

unprotected equipment.

4.3.3.2 ESD protection measures

When the equipment is housed in a metallic enclosure this itself can be used to guide the ESD current around the internal circuitry. Apertures or seams in the enclosure will act as high-impedance barriers to the current and transient fields will occur around them, so they must be minimized. All metallic covers and panels must be bonded together with a low impedance connection (<2.5mΩ at DC) in at least two places; long panel-to-panel "bonding" wires must be avoided since they radiate high fields during an ESD event. I/O cables and internal wiring may provide low-impedance paths for the current, in the same way as they are routes into and out of the equipment for common-mode RF interference. The best way to eliminate susceptibility of internal harnesses and cables is not to have any, through economical design of the board interconnections. External cables must have their shields well decoupled to the ground structure, following the rules in section 6.1.5, i.e. 360° bonding of cable screens to connector backshells and no pigtails [77].

Insulated enclosures make the control of ESD currents harder to achieve, and a well-designed and low inductance circuit ground is essential. But, if the enclosure can be designed to have no apertures which provide air gap paths to the interior then no discharge will be able to occur, provided the material's dielectric strength is high enough. You will still need to protect against the field of an indirect discharge, though.

4.3.4 Magnetic fields

Magnetic fields at low frequencies can induce interference voltages in closed wiring loops, their magnitude depending on the area that is intersected by the magnetic field. Non-toroidal mains transformers and switch-mode supply transformers are prolific sources of such fields and they will readily interfere with sensitive circuitry or components within the same equipment. Any other equipment needs to be immune to the proximity of such sources. Particular environments may result in high low-frequency or dc magnetic field strengths, such as electrolysis plant where very high currents are used, or certain medical apparatus. The voltage developed in a single turn loop is

$$V = A \cdot dB/dt \tag{4.8}$$

where A is the loop area in m^2 and
B is the flux density normal to the plane of the loop in Tesla

It is rare for such fields to affect digital or large signal analogue circuits, but they can be troublesome with low level circuits where the interference is within the operating bandwidth, such as audio or precision instrumentation. Specialized devices which are affected by magnetic fields, such as photomultiplier or cathode ray tubes, may also be susceptible.

4.3.4.1 Magnetic field screening

Conventional screening is ineffective against LF magnetic fields, because it relies on reflection rather than absorption of the field. Due to the low source impedance of magnetic fields reflection loss is low. Since it is only the component of flux normal to the loop which induces a voltage, changing the relative orientation of source and loop may be effective. LF magnetic shielding is only possible with materials which exhibit a high absorption loss such as steel, mu-metal or permalloy. As the frequency rises these materials lose their permeability and hence shielding efficiency, while non-magnetic materials such as copper or aluminium become more effective. Around 100kHz shielding efficiencies are about equal. Permeable metals are also saturated by high field strengths, and are prone to lose their permeability through handling.

4.3.5 Supply voltage fluctuations

Brown-outs (voltage droops) and interruptions are a feature of all mains distribution networks, and are usually due to fault clearing or load switching elsewhere in the system (Figure 4.21). Such events will not be perceived by the equipment if its input reservoir hold-up time is sufficient, but if this is not the case then restarts and output transients can be experienced. Typically, interruptions (as opposed to power cuts) can last for 10 – 500ms.

Load and line voltage fluctuations are maintained between +10% and –15% of the nominal line voltage in most industrialised countries. The majority of EEC countries are moving towards 230V +10% –6% or 230V +6% –10%. Slow changes in the voltage within these limits occur on a diurnal pattern as the load on the power system varies. The declared voltage does not include voltage drops within the customer's premises, and so you should design stabilised power supplies to meet at least the –15% limit. Dips exceeding 10% of nominal voltage occur up to 4 times per month for urban consumers and more frequently in rural areas where the supply is via overhead lines [44]. Note that much wider voltage (and frequency) fluctuations and more frequent interruptions are common in those countries which do not have a well-developed supply network. They are also common on supplies which are derived from small generators.

Harmonic distortion of the supply voltage is a function of loads which draw highly distorted current waveforms such as those discussed in section 4.2.3.2. Most electronic power supplies should be immune to such distortion, although it can have severe effects at high levels on power factor correction capacitors, motors and transformers, and may also interfere with audio systems.

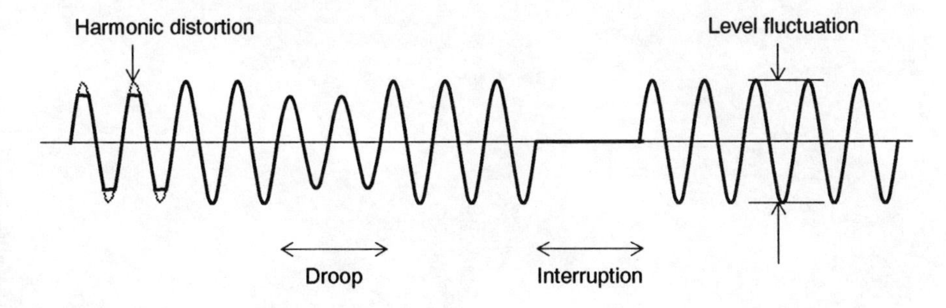

Figure 4.21 Mains supply fluctuations

Chapter 5

Circuits, layout and grounding

Designing for good EMC starts from the principle of controlling the flow of interference into and out of the equipment. You must assume that interference will occur to and will be generated by any product which includes active electronic devices. To improve the electromagnetic compatibility of the product you place barriers and route currents such that incoming interference is diverted or absorbed before it enters the circuit, and outgoing interference is diverted or absorbed before it leaves the circuit.

You can conceive the EMC control measures as applying at three levels, primary, secondary and tertiary, as shown in Figure 5.1.

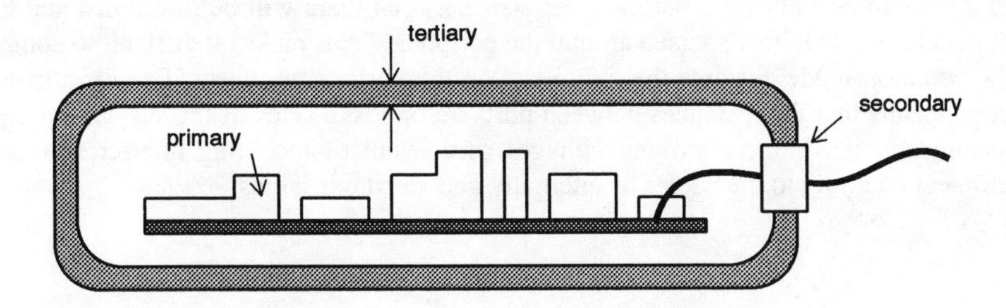

Figure 5.1 EMC control measures

Control at the primary level involves circuit design measures such as decoupling, balanced configurations, bandwidth and speed limitation, and also board layout and grounding. For some low-performance circuits, and especially those which have no connecting cables, such measures may be sufficient in themselves. At the secondary level you must always consider the interface between the internal circuitry and external cables. This is invariably a major route for interference in both directions, and for some products (particularly where the circuit design has been frozen) all the control may have to be applied by filtering at these interfaces. Choice and mounting of connectors forms an important part of this exercise.

Full shielding (the tertiary level) is an expensive choice to make and you should only choose this option when all other measures have been applied. But since it is difficult or impossible to predict the effectiveness of primary measures in advance, it is wise to allow for the possibility of being forced to shield the enclosure. This means adapting the mechanical design so that a metal case could be used, or if a moulded enclosure is essential, you should ensure that apertures and joints in the mouldings can be adequately bonded at RF, that ground connections can be made at the appropriate places and that the moulding design allows for easy application of a conductive coating.

5.1 Layout and Grounding

The most cost-effective approach is to consider the equipment's layout and ground regime at the beginning. No unit cost is added by a designed-in ground system. 90% of post-design EMC problems are due to inadequate layout or grounding: a well-designed layout and ground system can offer both improved immunity and protection against emissions, while a poorly designed one may well be a conduit for emissions and incoming interference. The most important principles are

- partition the system to allow control of interference currents
- consider ground as a path for current flow, both of interference into the equipment and conducted out from it; this means both careful placement of grounding points, and minimizing ground impedance
- minimize radiated emissions from, and susceptibility of current loops by careful layout of high di/dt loop areas

5.1.1 System partitioning

The first design step is to partition the system. A poorly partitioned, or non-partitioned system (Figure 5.2) may have its component sub-systems separated into different areas of the board or enclosure, but the interfaces between them will be ill-defined and the external ports will be dispersed around the periphery. This makes it difficult to control the common-mode currents that will exist on the various interfaces. Dispersal of the ports means that the distances between ports on opposite sides of the system is large, leading to high induced ground voltages as a result of incoming interference, and efficient coupling to the cables of internally generated emissions.

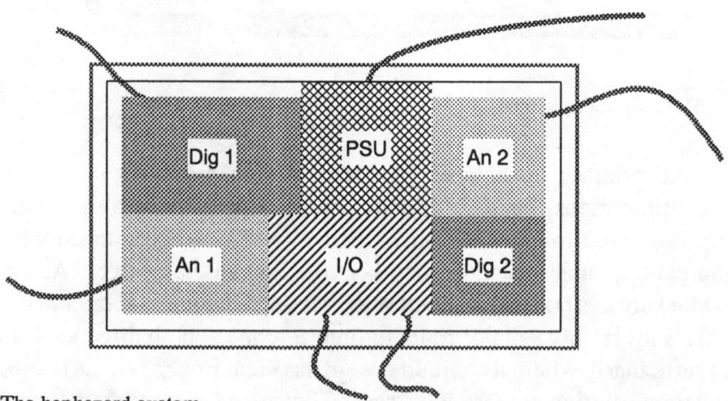

Figure 5.2 The haphazard system

Usually the only way to control emissions from and immunity of such a system is by placing an overall shield around it and filtering each interface. In many cases it will be difficult or impossible to maintain integrity of the shield and still permit proper operation – the necessary apertures and access points will preclude effective attenuation through the shield.

5.1.1.1 The partitioned system

Partitioning separates the system into critical and non-critical sections from the point of view of EMC. Critical sections are those which contain radiating sources such as

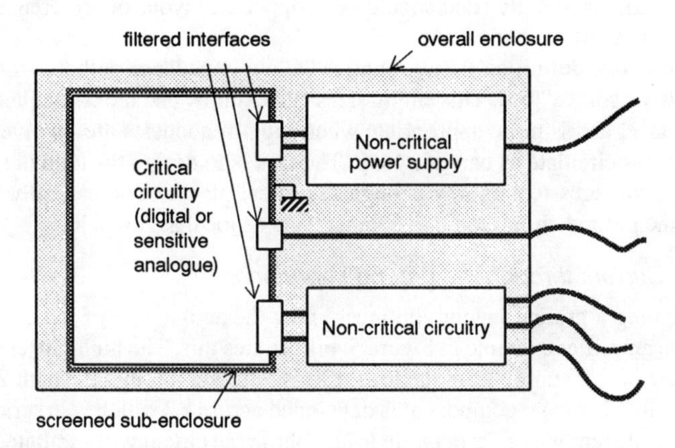

Figure 5.3 System partitioning

microprocessor logic or video circuitry, or which are particularly susceptible to imported interference: microprocessor circuitry and low-level analogue circuits. Non-critical sections are those whose signal levels, bandwidths and circuit functions are such that they are not susceptible to interference nor capable of causing it: non-clocked logic, linear power supplies and power amplifier stages are typical examples. Figure 5.3 shows this method of separation.

Control of critical sections

Critical sections can then be enclosed in a shielded enclosure into and out of which all external connections are carefully controlled. This enclosure may encase the whole product or only a portion of it, depending on the nature of the circuits: your major design goal should be to minimize the number of controlled interfaces, and to concentrate them physically close together. Each interface that needs to be filtered or requires screened cabling adds unit cost to the product. A system with no electrical interface ports – such as a pocket calculator or infra-red remote controller – represents an ideal case from the EMC point of view.

Note that the shield acts both as a barrier to radiated interference and as a reference point for ground return currents. In many cases, particularly where a full ground plane pcb construction is used, the latter is the more important function and it may be possible to do without an enclosing shield.

5.1.2 Grounding

Once the system has been properly partitioned, you can then ensure that it is properly grounded. There are two accepted purposes for grounding: one is to provide a route (the "safety earth") for hazardous fault currents, and the other is to give a reference for external connections to the system. The classical definition of a ground is "an equipotential point or plane which serves as a reference for a circuit or system". Unfortunately this definition is meaningless in the presence of ground current flow. Even where signal currents are negligible, induced ground currents due to environmental magnetic or electric fields will cause shifts in ground potential. A good grounding system will minimize these potential differences by comparison with the circuit operating levels, but it cannot eliminate them. It has been suggested that the term "ground" as conventionally used should be dropped in favour of "reference point" to make the purpose of the node clear.

An alternative definition for a ground is "a low impedance path by which current can return to its source"[63]. This emphasizes current flow and the consequent need for low impedance, and is more appropriate when high frequencies are involved. Ground currents always circulate as part of a loop. The task is to design the loop in such a way that induced voltages remain low enough at critical places. You can only do this by designing the ground circuit to be as compact and as local as possible.

5.1.2.1 Current through the ground impedance

When designing a ground layout you must know the actual path of the ground return current. The amplifier example in Figure 5.4 illustrates this. The high-current output ΔI returns to the power supply from the load; if it is returned through the path Z1–Z2–Z3 then an unwanted voltage component is developed across Z2 which is in series with the input V_S, and depending on its magnitude and phase the circuit will oscillate. This is an instance of common impedance coupling, as was covered in section 4.1.1.

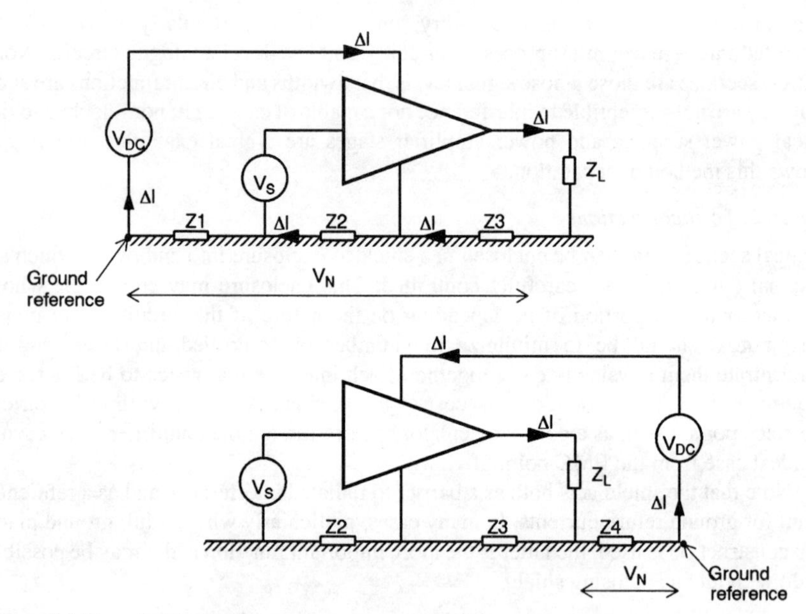

Figure 5.4 Ground current paths

A simple reconnection of the return path to Z4 eliminates the common impedance. For EMC purposes, instability is not usually the problem; rather it is the interference voltages V_N which are developed across the impedances that create emission or susceptibility problems. At high frequencies (above a few kHz) or high rates of change of current the impedance of any connection is primarily inductive and increases with frequency ($V = -L \cdot di/dt$), hence ground noise increases in seriousness as the frequency rises.

5.1.3 Ground systems

Ignoring for now the need for a safety earth, the grounding system as intended for a circuit reference can be configured as single point, multi-point or as a hybrid of these two.

5.1.3.1 Single point

The single point grounding system (Figure 5.5(a)) is conceptually the simplest, and it eliminates common impedance ground coupling and low frequency ground loops. Each circuit module has its own connection to a single ground, and each rack or sub-unit has one bond to the chassis. Any currents flowing in the rest of the ground network do not couple into the circuit. This system works well up to frequencies in the MHz region, but the distances involved in each ground connection mean that common mode potentials between circuits begin to develop as the frequency is increased. At distances greater than a quarter wavelength, the circuits are effectively isolated from each other.

A modification of the single point system (Figure 5.5(b)) ties together those circuit modules with similar characteristics and takes each common point to the single ground. This allows a degree of common impedance coupling between those circuits where it won't be a problem, and at the same time allows grounding of high frequency circuits to remain local. The noisiest circuits are closest to the common point in order to

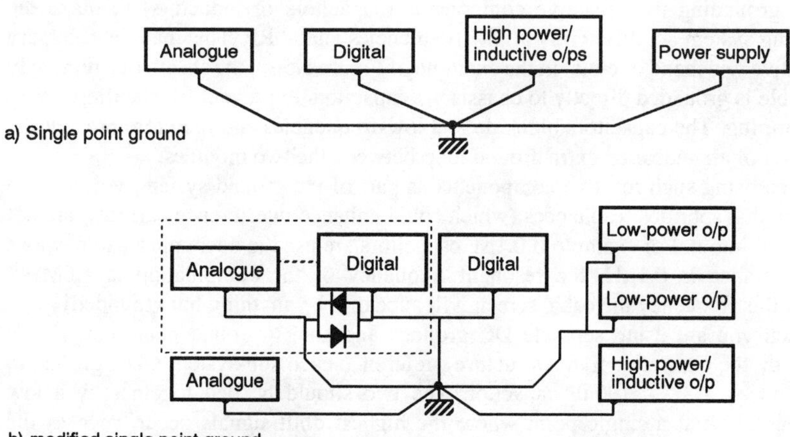

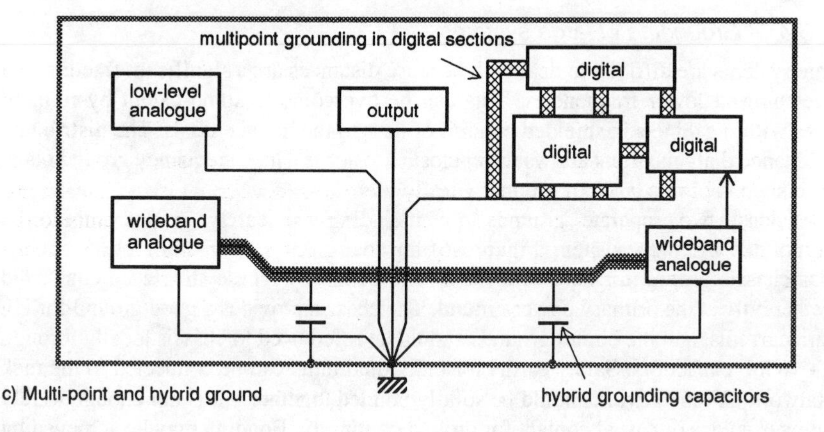

Figure 5.5 Grounding systems

minimize the effect of the common impedance. When a single module has more than one ground, these should be tied together with back-to-back diodes to prevent damage when the circuit is disconnected.

5.1.3.2 *Multi-point*

Hybrid and multi-point grounding (Figure 5.5(c)) can overcome the RF problems associated with pure single point systems. Multi-point grounding is necessary for digital and large signal high frequency systems. Modules and circuits are bonded together with many short ($< 0.1\lambda$) links to minimize ground-impedance-induced common mode voltages. Alternatively, many short links to a chassis,ground plane or other low impedance conductive body are made. This is not appropriate for sensitive analogue circuits, because the loops that are introduced are susceptible to magnetic field pick-up. It is very difficult to keep 50/60Hz interference out of such circuits. Circuits which operate at higher frequencies or levels are not susceptible to this interference. The multi-point subsystem can be brought down to a single point ground in the overall system.

5.1.3.3 *Hybrid*

Hybrid grounding uses reactive components (capacitors or inductors) to make the grounding system act differently at low frequencies and at RF. This may be necessary in sensitive wideband circuits. In the example of Figure 5.5(c), the sheath of a relatively long cable is grounded directly to chassis via capacitors to prevent RF standing waves from forming. The capacitors block dc and low-frequencies and therefore prevent the formation of an undesired extra ground loop between the two modules.

When using such reactive components as part of the ground system, you need to take care that spurious resonances (which could enhance interference currents) are not introduced into it. For example if 0.1μF capacitors are used to decouple a cable whose self-inductance is 0.1μH, the resonant frequency of the combination is 1.6MHz. Around this frequency the cable screen will appear to be anything but grounded!

When you are using separate DC grounds and an RF ground plane (such as is offered by the chassis or frame structure), reference each sub-system's DC ground to the frame by a 10 – 100nF capacitor. The two should be tied together by a low impedance link at a single point where the highest di/dt signals occur, such as the processor motherboard or the card cage backplane.

5.1.3.4 *Grounding of large systems*

Large systems are difficult to deal with because distances are a significant fraction of a wavelength at lower frequencies. This can be overcome to some extent by running cables within cabinets in shielded conduit or near to the metal chassis. The distributed capacitance that this offers allows the enclosure to act as a high frequency ground plane and to keep the impedance of ground wires low.

At least two separate grounds not including the safety ground should be incorporated within the system (Figure 5.6), an electronics ground return for the circuits and a chassis ground for hardware (racks and cabinets). These should be connected together only at the primary power ground. The chassis provides a good ground for HF return currents, and the circuit grounds should be referenced to chassis locally through 10 – 100nF capacitors. Safety earths for individual units can be connected to the rack metalwork. All metalwork should be solidly bonded together – it is not enough to rely on hinges, slides or casual contact for ground continuity. Bonding may be achieved by

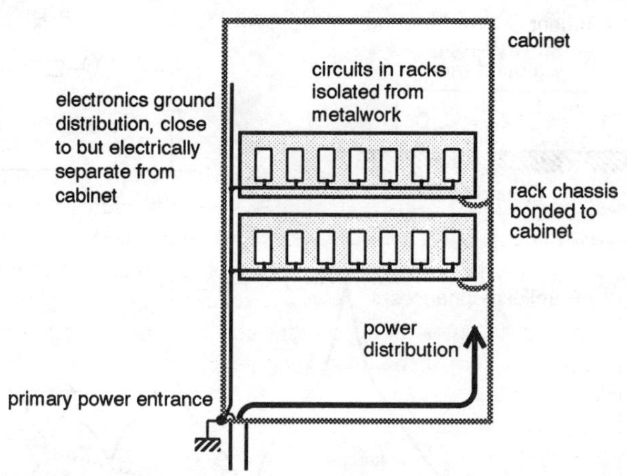

Figure 5.6 Grounding of a rack system

screw or rivet contact provided that steps are taken to ensure that paint or surface contamination does not interfere with the bond, but where panels or other structural components are removable then bonding should be maintained by a separate short, easily identified connecting strap.

The electronics ground can if necessary be subdivided further into "clean" and "dirty" ground returns for sensitive and noisy circuits, sometimes allocated as "signal" and "power" grounds. This allows low frequency single-point grounding either at each rack or at the overall cabinet ground.

5.1.3.5 The impedance of ground wires

When a grounding wire runs for some distance alongside a ground plane or chassis before being connected to it, it appears as a transmission line. This can be modelled as an LCR network with the L and C components determining the characteristic impedance Z_0 of the line (Figure 5.7). As the operating frequency rises, the inductive reactance exceeds the resistance of the wire and the impedance increases up to the first parallel resonant point. At this point the impedance seen at the end of the wire is high, typically hundreds of ohms (determined by the resistive loss in the circuit). After first resonance, the impedance for a lossless circuit follows the law

$$Z = Z_0 \cdot \tan(\omega \cdot x \cdot \sqrt{L/C}) \quad (5.1)$$

where x is the distance along the wire to the short

and successive series (low-impedance) and parallel (high-impedance) resonant frequencies are found. As the losses rise due to skin effect, so the resonant peaks and nulls become less pronounced. To stay well below the first resonance and hence remain an effective conductor, the ground wire should be less than 1/20th of the shortest operating wavelength.

5.1.3.6 The safety earth

From the foregoing discussion, you can see that the safety ground (the green and yellow wire) is not an RF ground at all. Many designers may argue that everything is connected to earth via the green and yellow wire, without appreciating that this wire has a high and

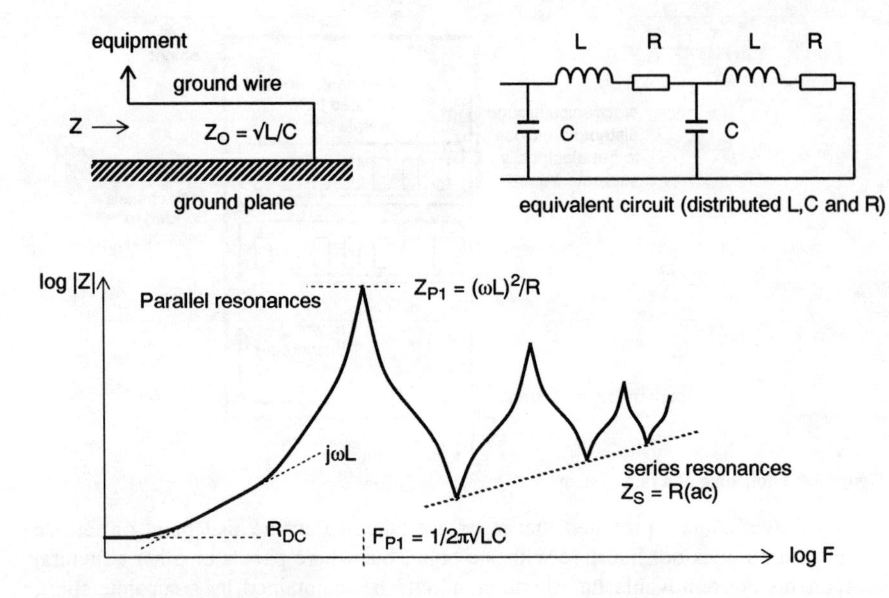

Figure 5.7 The impedance of long ground wires

variable impedance at RF. A good low impedance ground to a chassis, frame or plate is also necessary and in many cases must be provided *in parallel with* the safety earth. It may even be necessary for you to take the safety earth *out* of the circuit deliberately, by inserting a choke of the appropriate current rating in series with it.

5.1.3.7 Summary

For frequencies below 1MHz single-point grounding is possible and preferable. Above 10MHz a single-point ground system is not feasible because wire and track inductance raises the ground impedance unacceptably, and stray capacitance allows unintended ground return paths to exist. For high frequencies multi-point grounding to a low inductance ground plane or shield is essential. This creates ground loops which may be susceptible to magnetic field induction, so should be avoided or specially treated in a hybrid manner when used with very sensitive circuits.

For EMC purposes, *even a circuit which is only intended to operate at low frequencies must exhibit good immunity from RF interference.* This means that those aspects of its ground layout which are exposed to the interference – essentially all external interfaces – must be designed for multi-point grounding. At the bare minimum, some low-inductance ground plate or plane must be provided at the interfaces.

5.1.4 PCB layout

The way in which you design a printed circuit board makes a big difference to the overall EMC performance of the product which incorporates it. The principles outlined above must be carried through onto the pcb, particularly with regard to partitioning, interface layout and ground layout. This means that the circuit designer must exert tight control over the layout draughtsman, especially when CAD artwork is being produced. Conventional CAD layout software works on a node-by-node basis, which if allowed

Grounding principles

- All conductors have a finite impedance which increases with frequency
- Two physically separate ground points are not at the same potential unless no current flows between them
- At high frequencies there is no such thing as a single point ground

to will treat the entire ground system as one node, with disastrous results for the high frequency performance if left uncorrected.

The safest way to lay out a pcb (if a multilayer construction with power and ground planes is ruled out) is to start with the ground traces, manually if necessary, then to incorporate critical signals such as HF clocks or sensitive nodes which must be routed near to their ground returns, and then to track the rest of the circuitry at will. As much information should be provided with the circuit diagram as possible, to give the layout draughtsman the necessary guidance at the beginning. These notes should include

- physical partitioning of the functional sub-modules on the board
- positioning requirements of sensitive components and I/O ports
- marked up on the circuit diagram, the various *different* ground nodes that are being used, together with which connections to them must be regarded as critical
- where the various ground nodes may be commoned together, and where they must *not* be
- which signal tracks must be routed close to the ground tracks

5.1.4.1 Track impedance

Careful placement of ground connections goes a long way towards reducing the noise voltages that are developed across ground impedances. But on any non-trivial printed circuit board it is impractical to eliminate circulating ground currents entirely. The other aspect of ground design is to minimize the value of the ground impedance itself.

Track impedance is dominated by inductance at frequencies higher than a few kHz (Figure 5.8). You can reduce the inductance of a connection in two ways:

- minimizing the length of the conductor, and if possible increasing its width;
- running its return path parallel and close to it.

The inductance of a pcb track is primarily a function of its length, and only secondarily a function of its width. For a single track of width w and height h over a ground plane, equation (5.2) gives the inductance (further equations are given in Appendix C):

$$L = 0.005 \cdot \ln(2\pi \cdot h/w) \text{ microhenries per inch} \quad (5.2)$$

5.1.4.2 Minimizing ground impedance

The major factor in high-frequency impedance either of pc tracks or of wires is their length. Width, thickness or diameter are secondary factors. Because of the logarithmic relationship of inductance and width (equation (5.2)), doubling the width only produces a 75% decrease in inductance. Paralleling tracks will reduce the inductance *pro rata* provided that they are separated by enough distance to neutralise the effect of mutual inductance (see Figure 4.4). For narrow conductors spaced more than a centimetre apart, mutual inductance effects are negligible.

5.1.4.3 Gridded ground

The logical extension to paralleling ground tracks is to form the ground layout in a grid structure (Figure 5.9). This maximizes the number of different paths that ground return current can take and therefore minimizes the ground inductance for any given signal route. Such a structure is well suited to digital layout with multiple packages, when individual signal/return paths are too complex to define easily [38].

A wide ground track is preferred to narrow for minimum inductance, but even a narrow track linking two otherwise widely-separated points is better than none. The grid layout is best achieved by putting the grid structure down first, before the signal or power tracks are laid out. You can follow the X-Y routing system for double-sided boards, where the X-direction tracks are all laid on one side and the Y-direction tracks all on the other, provided the via hole impedance at junctions is minimized. Offensive (high di/dt) signal tracks can then be laid close to the ground tracks to keep the overall loop area small; this may call for extra ground tracking, which should be regarded as an acceptable overhead.

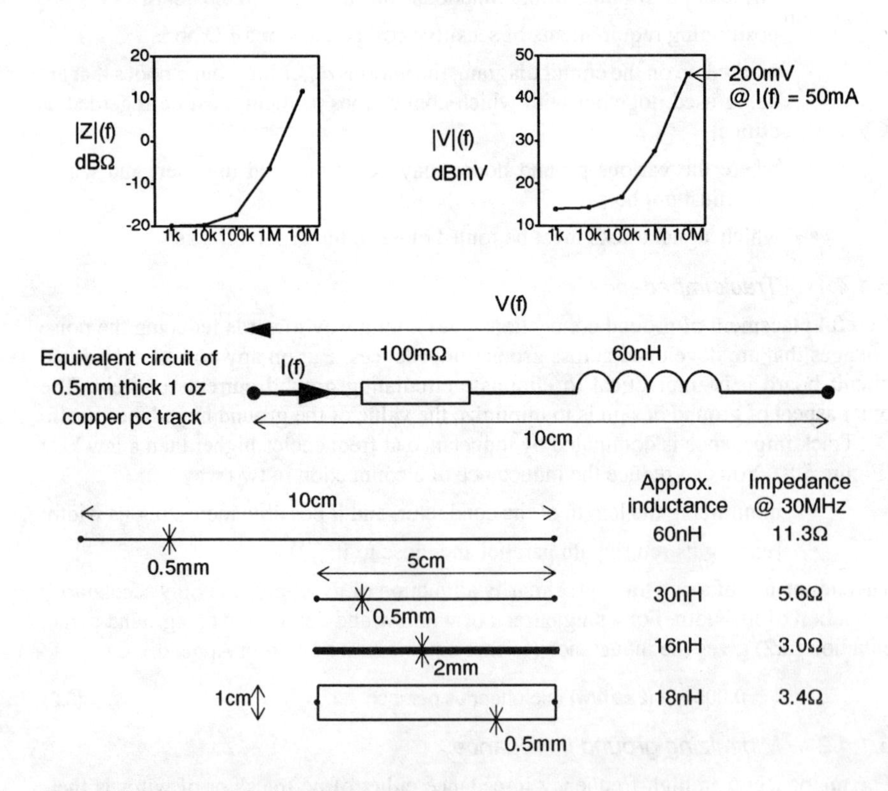

Figure 5.8 Impedance of printed circuit tracks

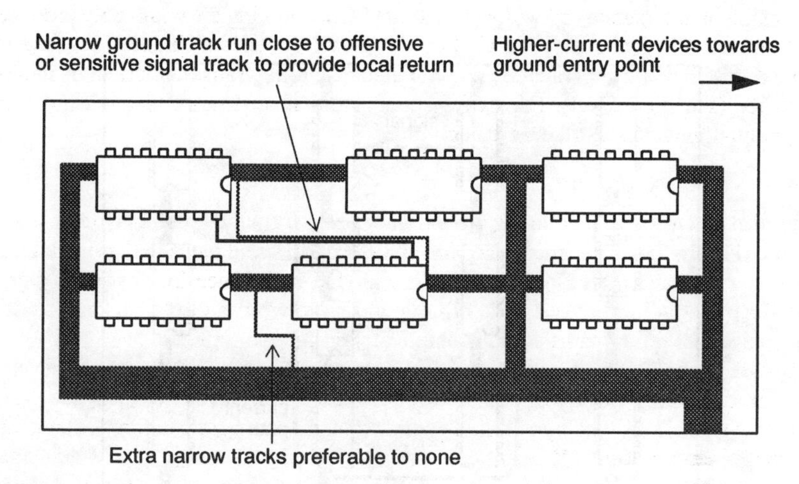

Figure 5.9 The gridded ground structure

Ground style versus circuit type

A gridded ground is not advisable for low frequency precision analogue circuits, because in these cases it is preferable to define the ground paths accurately to prevent common impedance coupling. Provided that the bandwidth of such circuits is low then high frequency noise due to ground inductance is less of a problem. For reduced ESD susceptibility, the circuit ground needs to remain stable during the ESD event. A low inductance ground network is essential, but this must also be coupled (by capacitors or directly) to a master reference ground structure.

The one type of ground configuration that you should not use for high speed logic is the "comb" style in which several ground spurs are run from one side of the board (Figure 5.10). Such a layout forces return currents to flow in a wide loop even if the signal track is short and direct, and contributes both to increased radiation coupling and to increased ground noise generation. The significant common ground impedance introduced between packages on the board may also cause circuit malfunction. The comb can easily be converted to a proper grid by adding bridging tracks at intervals across the spurs.

5.1.5 Ground plane

The limiting case of a gridded ground is when an infinite number of parallel paths are provided and the ground conductor is continuous, and it is then known as a ground plane. This is easy to realize with a multilayer board and offers the lowest possible ground path inductance. It is essential for RF circuits and digital circuits with high clock speeds, and offers the extra advantages of greater packing density and a defined characteristic impedance for all signal tracks [60]. A common four-layer configuration includes the power supply rail as a separate plane, which gives a low power-ground impedance at high frequencies.

Note that the main EMC purpose of a ground plane is to provide a low-impedance ground and power return path to minimize induced ground noise. Shielding effects on signal tracks are secondary and are in any case nullified by the component lead wires, when these stand proud of the board. There is little to be gained from having power and

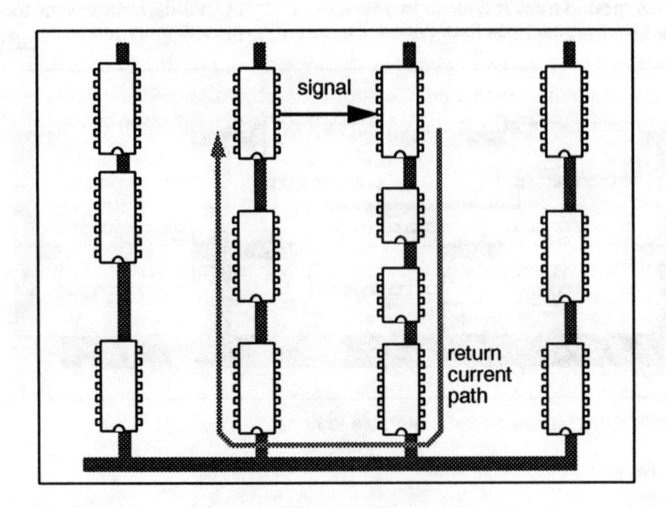

Figure 5.10 Undesirable: the comb ground structure

ground planes outside the signal planes on four-layer boards, especially considering the extra aggravation involved in testing, diagnostics and rework.

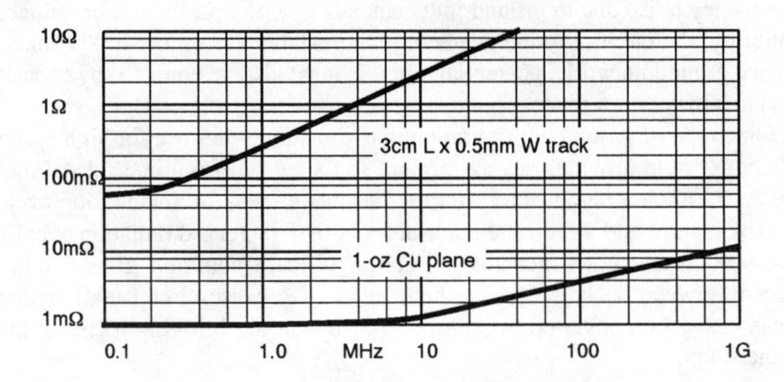

Figure 5.11 Impedance of ground plane versus track

Figure 5.11 compares the impedance between any two points (independent of spacing) on an infinite ground plane with the equivalent impedance of a short length of track. The impedance starts to rise at higher frequencies because the skin effect increases the effective resistance of the plane, but this effect follows the square root of frequency (10dB/decade) rather than the inductive wire impedance which is directly proportional to frequency (20dB/decade). For a finite ground plane, points in the middle will see the ideal impedance while points near the outside will see up to four times this value.

5.1.5.1 Ground plane on double-sided pcbs

A partial ground plane is also possible on double-sided pcbs. This is not achieved merely by filling all unused space with copper and connecting it to ground – since the purpose of the ground plane is to provide a low inductance ground path, it must be positioned under (or over) the tracks which need this low inductance return. At high frequencies, return current does not take the geographically shortest return path but will flow preferentially in the neighbourhood of its signal trace. This is because such a route encloses the least area and hence has the lowest overall inductance. Thus the use of an overall ground plane ensures that the optimum return path is always available, allowing the circuit to achieve minimum inductive loop area by its own devices [80].

A partial ground plane

Not all of the copper area of a complete ground plane needs to be used and it is possible to reduce the ground plane area by keeping it only under offending tracks. Figure 5.12 illustrates the development of the ground plane concept from the limiting case of two parallel identical tracks. To appreciate the factors which control inductance, remember

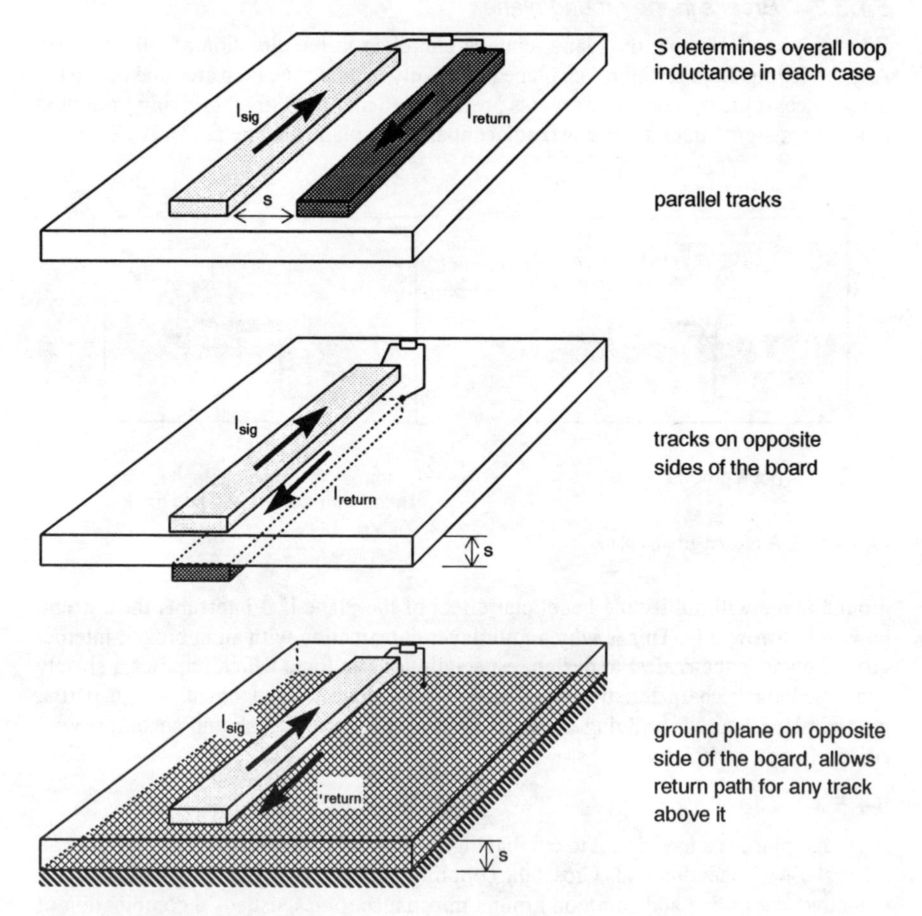

Figure 5.12 Return current paths

that the total loop inductance of two parallel tracks which are carrying current in *opposite* directions (signal and return) is given by equation (5.3):

$$L = L1 + L2 - 2M \tag{5.3}$$

where L1, L2 are the inductances of each track and
M is the mutual inductance between them

M is inversely proportional to the spacing of the tracks; if they were co-located it would be equal to L and the loop inductance would be zero. In contrast, the inductance of two identical tracks carrying current in the *same* direction is given by

$$L = (L + M)/2 \qquad (5.4)$$

so that a closer spacing of tracks increases the total inductance. Since the ground plane is carrying the return current for signal tracks above it, it should be kept as close as possible to the tracks to keep the loop inductance to a minimum. For a continuous ground plane this is set only by the thickness of the intervening board laminate.

5.1.5.2 Breaks in the ground plane

What *is* essential is that the plane remains unbroken in the direction of current flow. Any deviations from an unbroken plane effectively increase the loop area and hence the inductance. If breaks are necessary it is preferable to include a small bridging track next to a critical signal track to link two adjacent areas of plane (Figure 5.13). A slot in the

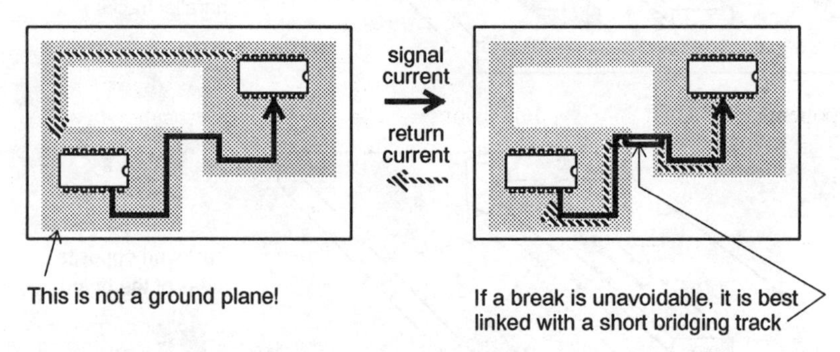
signal
current
return
current
This is not a ground plane!
If a break is unavoidable, it is best
linked with a short bridging track

Figure 5.13 A broken ground plane

ground plane will nullify the beneficial effect of the plane if it interrupts the current, however narrow it is. This is why a multilayer construction with an unbroken internal ground plane is the easiest to design, especially for fast logic which requires a closely controlled track characteristic impedance. Where double-sided board with a partial ground plane is used, bridging tracks as shown in Figure 5.13 should accompany all critical tracks, especially clocks.

5.1.5.3 Crosstalk

A ground plane is a useful tool to combat digital crosstalk, which is strictly speaking an internal EMC phenomenon. Crosstalk coupling between two tracks is mediated via inductive, capacitive and common ground impedance routes, usually a combination of all three (Figure 5.14). The effect of the ground plane is to significantly reduce the common ground impedance Z_G, by between 40 – 70dB. The actual improvement is not as great as this because the return current spreads out to some extent from directly underneath the signal track.

The ground plane may also reduce mutual inductance coupling by ensuring that the coupled current loops are not co-planar. Capacitive coupling will not be directly affected by the ground plane, but the lowered impedance of the line (equivalent to saying that C_{1G} and C_{2G} have been increased) will reduce capacitive crosstalk amplitude.

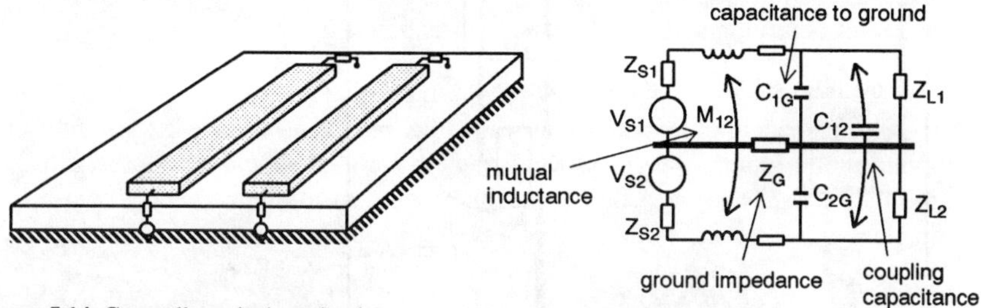

Figure 5.14 Crosstalk equivalent circuit

5.1.6 Loop area

The major advantage of a ground plane is that it allows for the minimum area of radiating loop. This ensures the minimum differential-mode emission from the pcb and also the minimum pickup of radiated fields. If you do not use a ground plane, it is still possible to ensure minimum loop area by keeping the tracks or leads that carry a high di/dt circuit, or a susceptible circuit, close to one another [64]. This is helped by using components with small physical dimensions, and by keeping the separation distance as small as possible consistent with layout rules. Such circuits (e.g. clocks or sensitive inputs) should not have their source and destination far apart. The same also applies to power rails. Transient power supply currents should be decoupled to ground close to their source, by several low-value capacitors well distributed around the board.

Case studies

Two examples of poor layout from the point of view of minimizing high di/dt loops are shown in Figure 5.15. The first concerns a 68HC11 single chip microprocessor whose E clock output at 2MHz was feeding a 74HC00 gate for timing purposes. Another output of the 74HC00 was fed back to a port input on the microprocessor. The two chips were positioned close together so that the signal tracks were relatively short (about 5cm). Unfortunately their 0V returns were connected to opposite ends of a long ground trace, so that the 2MHz squarewave return currents flowed around a loop that was virtually the entire area of the board! Making a simple link from point A to point B (one extra short track) reduced the emissions of higher order 2MHz harmonics by 15–20dB. A further improvement of several dB was obtained by re-laying the board with a true gridded ground system.

The second example is a small isolating switching supply using a power MOSFET running at 400kHz. Transition times were of the order of 10ns so the harmonic content of the switching waveform extended towards 100MHz. The electrolytic decoupling capacitor was positioned several cm away from the other components with the result that a large-area current loop existed between power and ground rails – even though the rail tracks were reasonably wide. Including a 47nF RF decoupling capacitor right next to the power transistor and transformer reduced the HF currents flowing in this loop to the extent that conducted emissions above 10MHz dropped by 20dB.

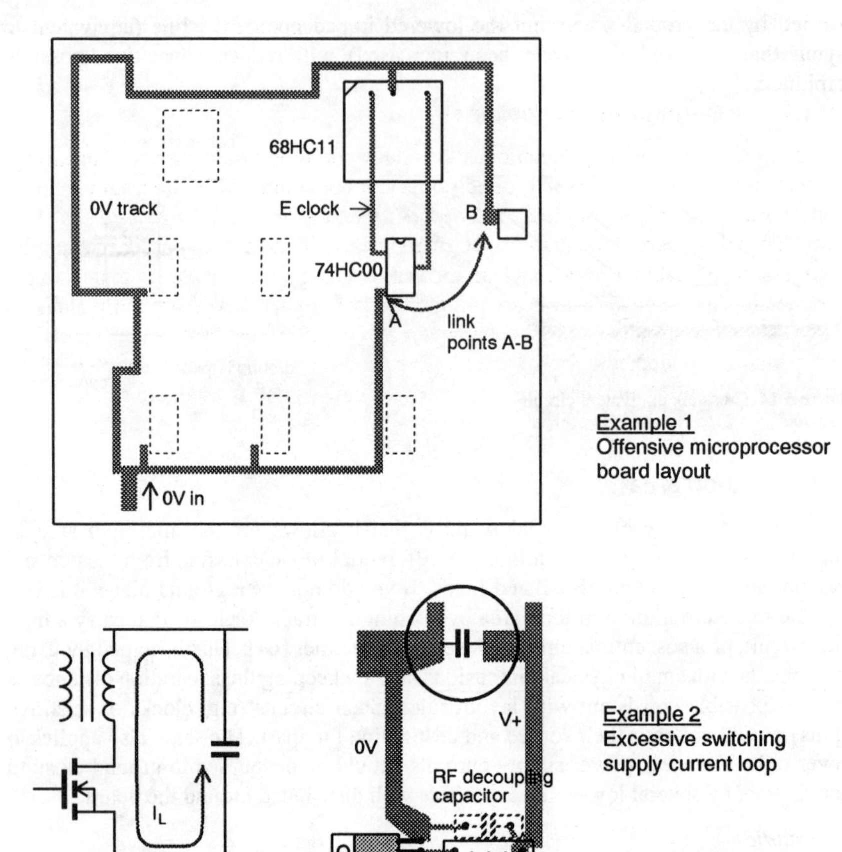

Figure 5.15 Loop area: case studies

5.1.6.1 The advantage of surface mount

Surface mount technology (SMT) offers smaller component sizes and therefore should give a reduction in interference coupling, since the overall circuit loop area can be smaller. This is in fact the case, but to take full advantage of SMT a multilayer board construction with ground plane is necessary. There is a slight improvement when a double-sided board is re-laid out to take SMT components, which is mainly due to shrinking the overall board size and reducing the length of individual tracks. The predominant radiation is from tracks rather than components.

But when a multilayer board is used, the circuit loop area is reduced to the track length times the track-to-ground plane height. Now, the dominant radiation is from the extra area introduced by the component leadouts. The reduction in this area afforded by SMT components is therefore worthwhile. For EMC purposes, SMT and multilayer groundplane construction are complementary.

A further advantage of surface mount is that, rather than taking advantage of the component size reduction to pack more functions into a given board area, you can reduce the board area needed for a given function. This allows you more room to define quiet I/O areas and to fit suppression and filtering components when these prove to be needed.

5.1.7 Configuring I/O grounds

Decoupling and shielding techniques to reduce common-mode currents appearing on cables both require a "clean" ground point, not contaminated by internally generated noise. *Filtering at high frequencies is next to useless without such a ground.* Unless you consider this as part of the layout specification early in the design phase, such a ground will not be available. Provide a clean ground by grouping all I/O leads in one area and connecting their shields and decoupling capacitors to a separate ground plane in this area. The clean ground can be a separate area of the pcb [65], or it can be a metal plate on which the connectors are mounted. The external ground (which may be only the mains safety earth) and the metal or metallized case, if one is used, are connected here as well, via a low-inductance link. Figure 5.16 shows a typical arrangement for a product with digital, analogue and interface sections.

This clean ground must only connect to the internal logic ground at one point. This prevents logic currents flowing through the clean ground plane and "contaminating" it. No other connections to the clean ground are allowed. As well as preventing common mode emissions, this layout also shunts incoming interference currents (transient or RF) to the clean ground and prevents them flowing through susceptible circuitry.

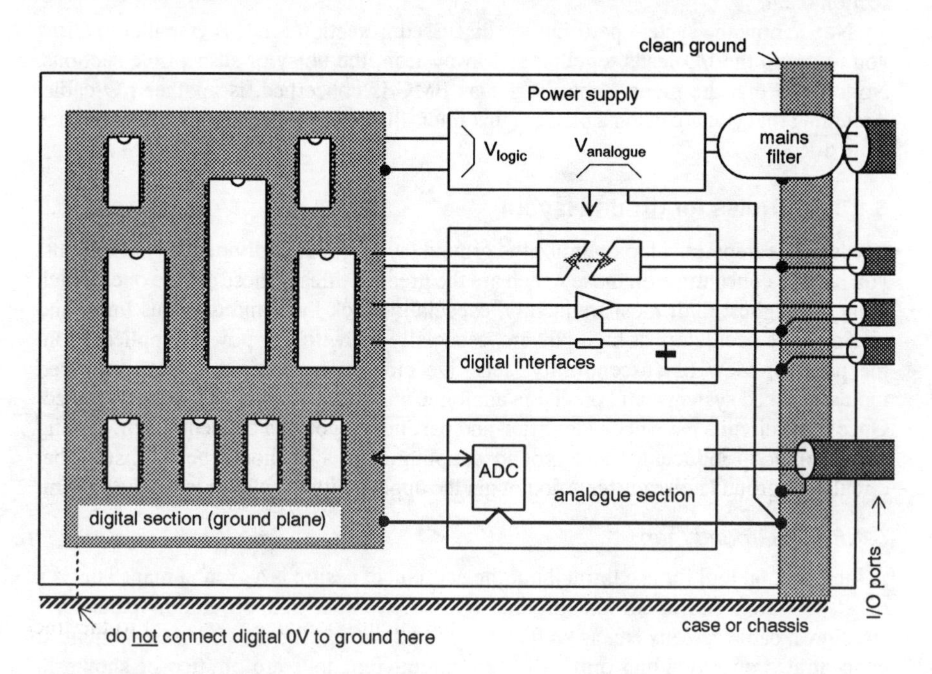

Figure 5.16 Grounding at the interfaces

If for other reasons it is essential to have leads interfacing with the unit or pcb at different places, you should still arrange to couple them all to a clean ground, that is one through which no circuit currents are flowing. In this case a chassis plate is mandatory.

For ESD protection the circuit ground *must* be referenced to the chassis ground.

This can be easily done by using a plated-through hole on the ground track and a metallic standoff spacer. If there has to be DC isolation between the two grounds at this point, use a 10 – 100nF RF (ceramic or polyester) capacitor. You can provide a clean I/O ground on plug-in rack mounting cards by using wiping finger style contacts to connect this ground track directly to the chassis.

5.1.7.1 Separate circuit grounds

Never extend a digital ground plane over an analogue section of the pcb as this will couple digital noise into the analogue circuitry. A single-point connection between digital and analogue grounds can be made at the system's analogue-to-digital converter. It is very important *not* to connect the digital cicuitry separately to an external ground [27]. If you do this, extra current paths are set up which allow digital circuit noise to circulate in the clean ground.

Interfaces directly to the digital circuitry (for instance, to a port input or output) should be buffered so that they do not need to be referenced to the digital 0V. The best interface is an opto-isolator or relay, but this is of course expensive. When you can't afford isolation, a separate buffer IC which can be referenced to the I/O ground is preferable; otherwise, buffer the port with a series resistor or choke and decouple the line *at the board interface* (not somewhere in the middle of the board) with a capacitor and/or a transient suppressor to the clean ground. More is said about I/O filtering in section 6.2.4.

Notice how the system partitioning, discussed in section 5.1.1, is essential to allow you to group the I/O leads together and away from the noisy or susceptible sections. Notice also that the mains cable, as far as EMC is concerned, is another I/O cable. Assuming that you are using a block mains filter, fit this to the "clean" ground reference plate directly.

5.1.8 Rules for ground layout

Because it is impractical to optimise the ground layout for all individual signal circuits, you have to concentrate on those which are the greatest threat. These are the ones which carry the highest di/dt most frequently, especially clock lines and data bus lines, and square-wave oscillators at high powers, especially in switching power supplies. From the point of view of susceptibility, sensitive circuits – particularly edge-triggered inputs, clocked systems, and precision analogue amplifiers – must be similarly treated. Once these circuits have been identified and partitioned you can concentrate on dealing with their loop inductance and ground coupling. The aim should be to ensure that circulating ground noise currents do not get the opportunity to enter or leave the system.

5.1.8.1 Ground map

A fundamental tool for use throughout the equipment design is a ground map. This is a diagram which shows all the ground reference points and grounding paths (via structures, cable screens etc. as well as tracks and wiring) for the whole equipment. It concentrates on grounding only; all other circuit functions are omitted or shown in block form. Its creation, maintenance and enforcement throughout the project design should be the responsibility of the EMC design authority.

5.2 Digital and analogue circuit design

Digital circuits are prolific generators of electromagnetic interference. High-frequency

Grounding rules

- identify the circuits of high di/dt (for emissions) –
  clocks, bus buffers/drivers, high-power oscillators
- identify sensitive circuits (for susceptibility) –
  low-level analogue, fast digital data
- minimize their ground inductance by –
  minimizing the length and enclosed area
- apply partitioning wherever possible
- ensure that internal and external ground noise cannot couple out of or into the system
- create, maintain and enforce a ground map

square-waves, rich in harmonics, are distributed throughout the system. The harmonic frequency components reach into the part of the spectrum where cable resonance effects are important. Analogue circuits are in general much quieter because high frequency squarewaves are not normally a feature. A major exception is wide bandwidth video circuits, which transmit broadband signals up to several MHz, or several tens of MHz for high resolution video. Any analogue design which includes a high frequency oscillator or other high di/dt circuits must follow HF design principles, especially with regard to ground layout.

Some low frequency amplifier circuits can oscillate in the MHz range, especially when driving a capacitive load, and this can cause unexpected emissions. The switching power supply is a serious cause of interference at low to medium frequencies since it is essentially a high-power squarewave oscillator.

Because the microprocessor is a state machine, processor-based circuits are prone to corruption by fast transients which can cause the execution of false states. Great care is necessary to prevent any clocked circuit (not just microprocessor-based) from being susceptible to incoming interference. Analogue signals are more affected by continuous interference, which is rectified by non-linear circuit elements and causes bias or signal level shifts. The immunity of analogue circuits is improved by minimizing amplifier bandwidth, maximizing the signal level, using balanced configurations and electrically isolating I/O that will be connected to “dirty” external circuits.

5.2.1 The Fourier spectrum

5.2.1.1 The time domain and the frequency domain

Basic to an understanding of why switching circuits cause interference is the concept of the time domain/frequency domain transform. Most circuit designers are used to working with waveforms in the time domain, as viewed on an oscilloscope, but any repeating waveform can also be represented in the frequency domain, for which the basic measuring and display instrument is the spectrum analyser (section 3.1.1.2). Whereas the oscilloscope shows a segment of the waveform displayed against time, the spectrum analyser will show the same waveform displayed against frequency. Thus the relative amplitudes of different frequency components of the signal are instantly seen.

The mathematical tool which allows you to analyse a known time-domain waveform in the frequency domain is called the Fourier transform. The necessary equations for the Fourier transform are covered in Appendix C. Figure 5.17 shows the spectral amplitude compositions of various types of waveform (phase relationships are rarely of any interest for EMC purposes). The sine wave has only a single component at its fundamental frequency. A square wave with infinitely fast rise and fall times has a series of odd harmonics (multiples of the fundamental frequency) extending to infinity. A sawtooth contains both even and odd harmonics.

Switching waveforms can be represented as trapezoidal; a digital clock waveform is normally a square wave with defined rise and fall times. The harmonic amplitude content of a trapezoid decreases from the fundamental at a rate of 20dB per decade until a breakpoint is reached at $1/\pi t_r$, after which it decreases at 40dB/decade (Figure 5.18(a)). Of related interest is the differentiated trapezoid, which is an impulse with finite rise and fall times. This has the same spectrum as the trapezoid at higher frequencies, but the amplitude of the fundamental and lower order harmonics is reduced and flat with frequency. (This property is intuitively obvious as a differentiator has a rising frequency response of +20dB/decade.) Reducing the trapezoid's duty cycle from 50% has the same effect of decreasing the fundamental and low frequency harmonic content.

Asymmetrical slew rates generate even as well as odd harmonics. This feature is important, since differences between high- and low-level output drive and load currents mean that most logic circuits exhibit different rise and fall times, and it explains the presence and often preponderance of even harmonics at the higher frequencies.

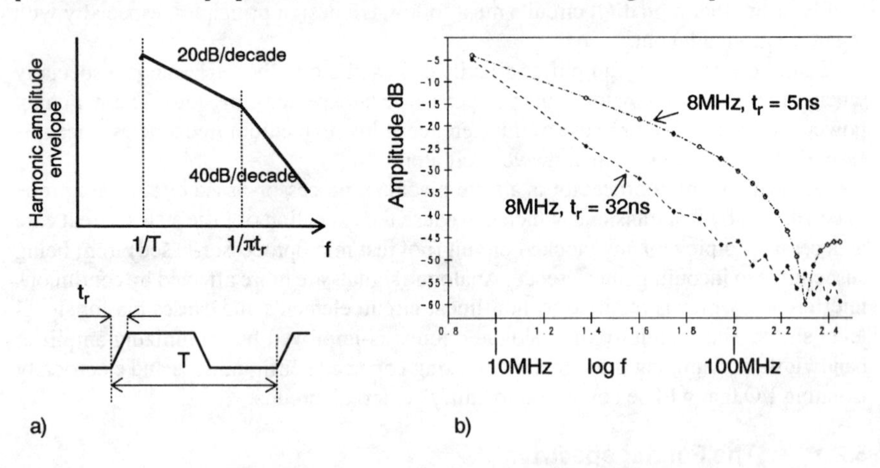

Figure 5.18 Harmonic envelope of a trapezoid

5.2.1.2 Choice of logic family

The damage as far as emissions are concerned is done by switching edges which have a fast rise or fall time (note that this is not the same as transition time and is rarely specified in data sheets; where it is, it is usually a maximum figure). Using the slowest risetime compatible with reliable operation will minimise the amplitude of the higher-order harmonics where radiation is more efficient. Figure 5.18(b) shows the calculated harmonic amplitudes for an 8MHz clock with risetimes of 5ns and 32ns. An

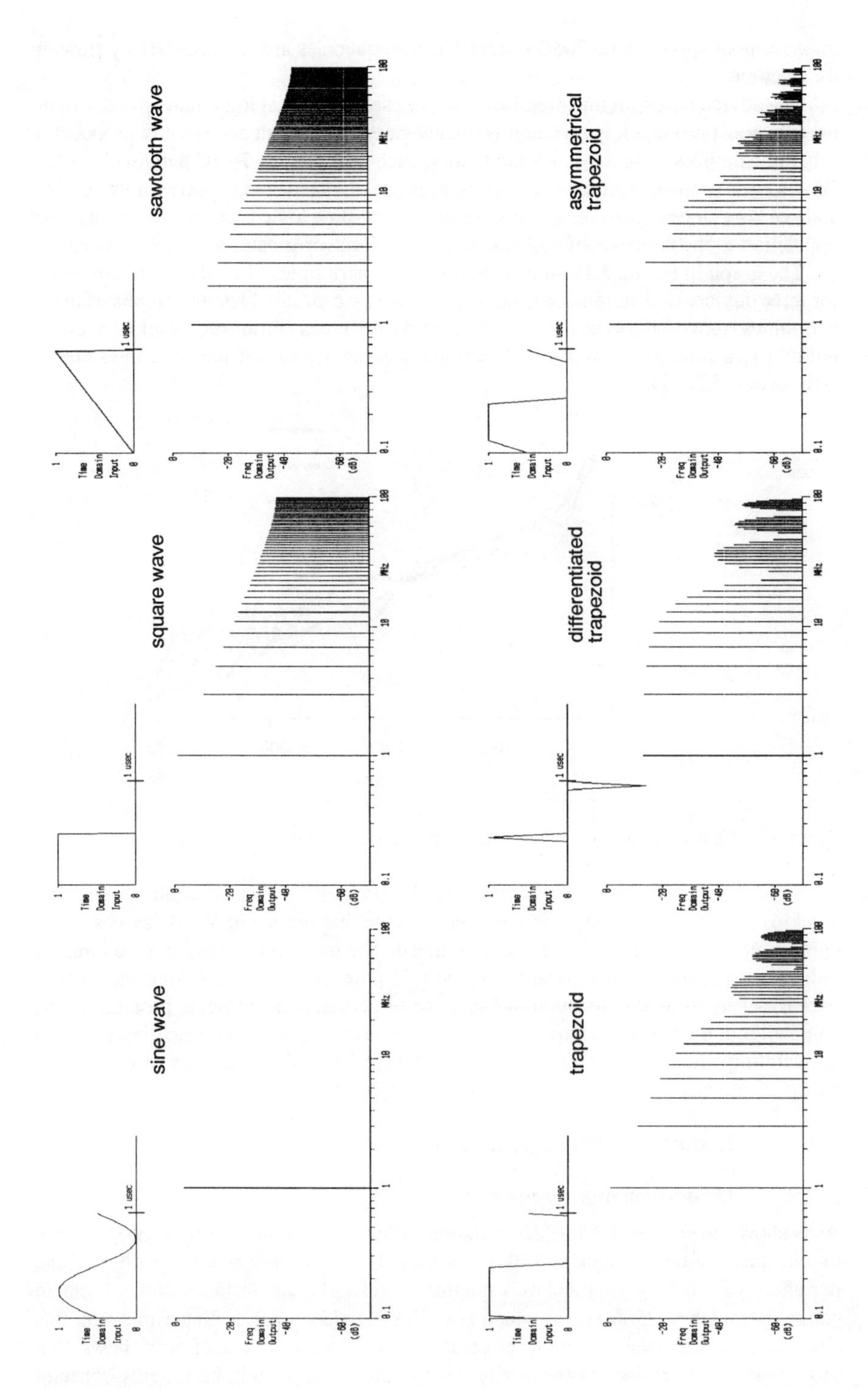

Figure 5.17 Frequency spectra for various waveforms

improvement approaching 20dB is possible at frequencies around 100MHz by slowing the risetime.

The advice based on this premise must be, use the slowest logic family that will do the job; don't use fast logic when it is unnecessary. Treat with caution any proposal to substitute devices from a faster logic family, such as replacing 74HC parts with 74AC. Where parts of the circuit must operate at high speed, use fast logic only for those parts and keep the clock signals local. This preference for slow logic is unfortunately in direct opposition to the demands of software engineers for ever-greater processing speeds.

The graph in Figure 5.19 shows the measured harmonics of a 10MHz square wave for three devices of different logic families in the same circuit. Note the emphasis in the harmonics above 200MHz for the 74AC and 74F types. From the point of view of immunity, a slow logic family will respond less readily to fast transient interference (see section 5.2.6.1).

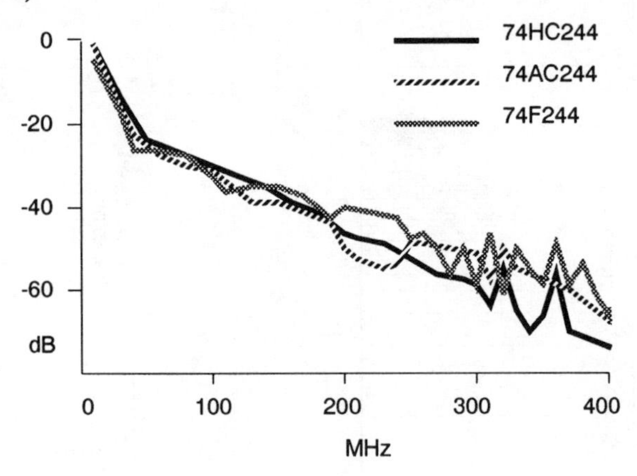

Figure 5.19 Comparison of harmonic spectra of different logic families

Some IC manufacturers are addressing the problem of RF emissions at the chip level [61]. By careful attention to the internal switching regime of VLSI devices, noise currents appearing at the pins can be minimized. The transition times can be optimized rather than minimized for a given application [39]. Revised package design and smaller packages can allow the decoupling capacitor to be placed as close as possible to the chip, without the internal leadframe's inductance negating its effect; also, the reduction in operating silicon area gained from shrinking silicon design rules can be used to put a respectable-sized decoupling capacitor (say 1nF) actually on the silicon.

5.2.2 Radiation from logic circuits

5.2.2.1 Differential mode radiation

As is shown in section 4.2.1.1, the radiation efficiency of a small loop is proportional to the square of the frequency (+40dB/decade). This relationship holds good until the periphery of the loop approaches a quarter wavelength, for instance about 15cm in epoxy-glass pcb at 250MHz, at which point the efficiency peaks. Superimposing this characteristic onto the harmonic envelope of a trapezoidal waveform, shows that differential-mode emissions (primarily due to current loops) will be roughly constant with frequency (Figure 5.20(a)) above a breakpoint determined by the risetime [10].

The actual radiated emission envelope at 10m can be derived from equation (4.4) provided the peak-to-peak squarewave current, risetime and fundamental frequency are known. The Fourier coefficient at the fundamental frequency F1 is 0.64, therefore the emission at F1 will be

$$E = 20\log_{10} . [119 . 10^{-6} (f^2 . A . I_{pk})] \text{ dB}\mu\text{V/m} \quad (5.5)$$

from which the +20dB/decade line to the breakpoint at $1/\pi t_r$ is drawn. In fact, by

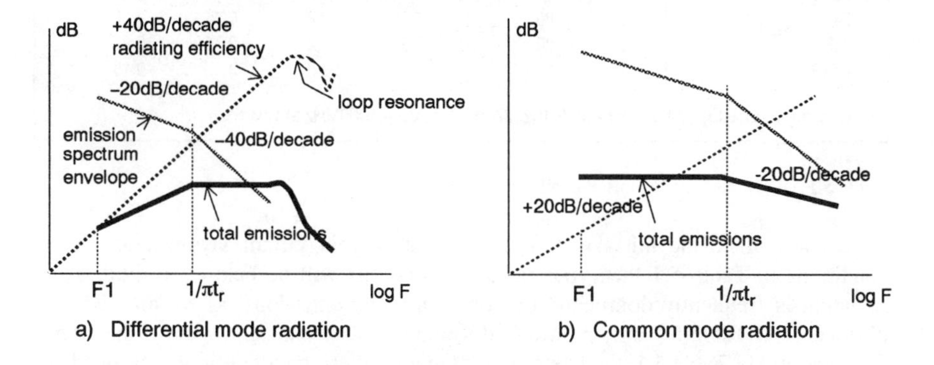

Figure 5.20 Emissions from digital trapezoid waves via different paths

combining the known rise and fall times and transient output current capability for a given logic family with the trapezoid Fourier spectrum at various fundamental frequencies, the maximum radiated emission can be calculated for different loop areas. If these figures are compared with the EN Class B radiated emissions limit (30dBμV/m at 10m up to 230MHz), a table of maximum allowable loop area for various logic families and clock frequencies can be derived. Table 5.1 shows such a table.

The ΔI figure in Table 5.1 is the dynamic switching current that can be supplied by the device to charge or discharge the node capacitance. It is not the same as the steady-state output high or low load current capability. Some manufacturers' data will give these figures as part of the overall family characteristics. The same applies to t_r and t_f, which are typical figures, and are not the same as maximum propagation delay specifications.

Layout and construction implications

The implication of these figures is that for clock frequencies above 30MHz, or for fast logic families (AC, AS or F), a ground plane layout is essential as the loop area restrictions cannot be met in any other way. Even this is insufficient if you are using fast logic at clock frequencies above 30MHz. The loop area introduced by the device package dimensions exceeds the allowed limit and extra measures (shielding and filtering) are unavoidable. This information can be useful in the system definition stages of a design, when the extra costs inherent in choosing a higher clock speed (versus, for example, multiple processors running at lower speeds) can be estimated.

Table 5.1 applies to a single radiating loop. Usually, only a few tracks dominate the radiated emissions profile. Radiation from these loops can be added on a root mean square basis, so that for n similar loops the emission is proportional to $\sqrt{n}$. If the loops carry signals at different frequencies then their emissions will not add.

Logic family	t_r/t_f ns	ΔI mA	Loop area cm^2 at clock frequency 4MHz	10MHz	30MHz	100MHz
4000B CMOS @ 5V	40	6	1000	400	-	-
74HC	6	20	45	18	6	-
74LS	6	50	18	7.2	2.4	-
74ALS	3.5	50	10	4	1.4	0.4
74AC	3	80	5.5	2.2	0.75	0.25
74F	3	80	5.5	2.2	0.75	0.25
74AS	1.4	120	2	0.8	3	0.15

Loop area for 30dBμV/m 30MHz – 230MHz, 37dBμV/m 230 – 1000MHz at 10m

Table 5.1 Differential mode emission: allowable loop area

Do not make the mistake of thinking that if your circuit layout satisfies the conditions in Table 5.1 then your radiated emissions will be below the limit. Total radiation is frequently dominated by common-mode emissions, as we are about to discover, and Table 5.1 only relates to differential-mode emissions. But if the circuit does *not* satisfy Table 5.1, then extra shielding and filtering will definitely be needed.

5.2.2.2 Common mode radiation

Common mode radiation which is due mainly to cables and large metallic structures increases at a rate linearly proportional to frequency, as shown in equation (4.5). Thus combining the harmonic spectrum and the radiating efficiency as in section 5.2.2.1 gives a radiated field which is constant up to the $1/\pi t_r$ breakpoint and then decreases at a rate of 20dB/decade (Figure 5.20(b)). It might appear from this that this coupling mode is less important. This comparison is misleading for two reasons:

- cable radiation is much more effective than from a small loop, and so a smaller common-mode current (of the order of microamps) is needed for the same field strength;
- cable resonance usually falls within the range 30 – 100MHz, and radiation is enhanced over that of the short cable model. In fact, common-mode coupling is usually the major source of radiated emissions.

A similar calculation to that done for differential mode can be done for cable radiation on the basis of the model shown in Figure 5.21. This assumes that the cable is driven by a common-mode voltage developed across a ground track which forms part of a logic circuit. The ground track carries the current ΔI which is separated into its frequency components by Fourier analysis, and this current then generates a noise voltage differential V_N of $\Delta I \cdot j\omega \cdot L$ between the ground reference (assumed to be at one end of the track) and the cable connection (assumed to be at the other). A factor of –20dB is allowed for lossy coupling to the ground reference. The cable impedance is assumed to be a resistive 150Ω and constant with frequency – this is a crude approximation generally borne out in practice.

Track length implications

The inductance L is crucial to the level of noise that is emitted. In the model, it is calculated from the length of a 0.5mm wide track separated from its signal track by

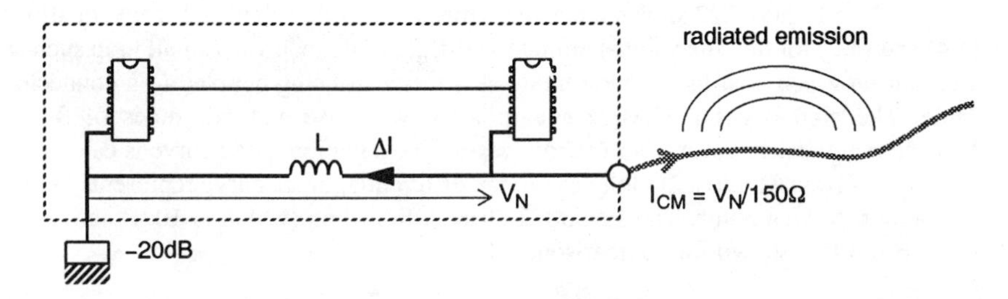

Figure 5.21 Common mode emission model

0.5mm, so that mutual inductance cancellation reduces the overall inductance. Table 5.2 tabulates the resulting allowable track lengths versus clock frequency and logic family as before, for a radiated field strength corresponding to the EN Class B limits.

Logic family	t_r/t_f ns	ΔI mA	Track length cm at clock frequency 4MHz	10MHz	30MHz	100MHz
4000B CMOS @ 5V	40	6	180	75	-	-
74HC	6	20	8.5	3	1	-
74LS	6	50	3.25	1.3	0.45	-
74ALS	3.5	50	1.9	0.75	0.25	0.08
74AC	3	80	1.0	0.4	0.14	0.05
74F	3	80	1.0	0.4	0.14	0.05
74AS	1.4	120	0.4	0.15	0.05	-

Allowable track length for 30dBμV/m 30MHz – 230MHz, 37dBμV/m 230 – 1000MHz at 10m; cable length = 1m; layout: parallel 0.5mm tracks 0.5mm apart (2.8nH/cm)

Table 5.2 Common mode emission: allowable track length

This model should not be taken too seriously for prediction purposes. Too many factors have been simplified: cable resonance and impedance variations with frequency and layout, track and circuit resonance and self-capacitance, and resonance and variability of the coupling path to ground have all been omitted. The purpose of the model is to demonstrate that logic circuit emissions are normally dominated by common mode factors. Common mode currents can be combated by

- ensuring that logic currents do not flow between the ground reference point and the point of connection to external cables;
- filtering all cable interfaces to a "clean" ground;
- screening cables with the screen connection made to a "clean" ground;
- minimizing ground noise voltages by using low-inductance ground layout or, preferably, a ground plane.

Table 5.2 shows that the the maximum allowable track length for the higher frequencies and faster logic families is impracticable (fractions of a millimetre!). Therefore one or a combination of the above techniques will be essential to bring such circuits into compliance.

5.2.2.3 Comparison of differential and common mode

The graph in Figure 5.22 shows the actual emission profile calculated from equations (4.4) and (4.5) for the same signal emitted in differential mode via a small loop such as a circuit on a pcb, and in common mode as a result of being coupled to a connected cable. The signal was a 12MHz square wave with rise and fall times of 3.5ns. Measurement distance was 10m for both cases. The common mode curve is calculated for 0.1mA (100μA) into a 2m long cable; the differential mode curve represents 20mA into a loop of area 5cm^2. The EN55022 class A limit line (which is 10dB above the Class B limit) is shown for comparison.

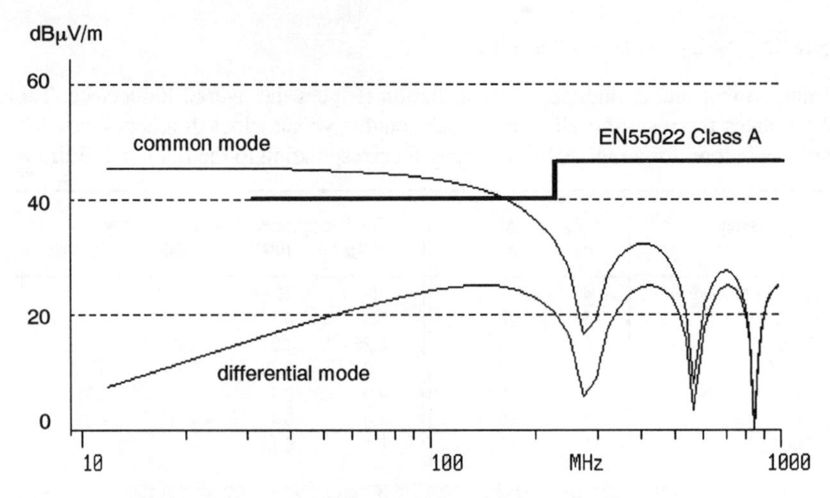

Figure 5.22 Comparison of common mode and differential mode emissions

5.2.2.4 Clock and broadband radiation

The main source of radiation in digital circuits is the processor clock (or clocks) and its harmonics. All the energy in these signals is concentrated at a few specific frequencies, with the result that clock signal levels are 10–20dB higher than the rest of the digital circuit radiation. Since the commercial radiated emissions standards do not distinguish between narrowband and broadband, these narrowband emissions should be minimized first, by proper layout and grounding of clock lines. Then pay attention to other broadband sources, especially data/address buses and backplanes, and video or high-speed data links.

Backplanes

Buses which drive several devices or backplanes which drive several boards carry much higher switching currents (because of the extra load capacitance) than circuits which are compact and/or lightly loaded. Products which incorporate a backplane are more prone to high radiated emissions. A high-speed backplane should always use a multilayer board with a ground plane, and daughter board connectors should include a ground pin for every high-speed clock, data or address pin (Figure 5.23). If this is impractical, multiple distributed ground returns can be used to minimize loop areas. The least significant data/address bit usually has the highest frequency component of a bus and should be run closest to its ground return. Clock distribution tracks must *always* have

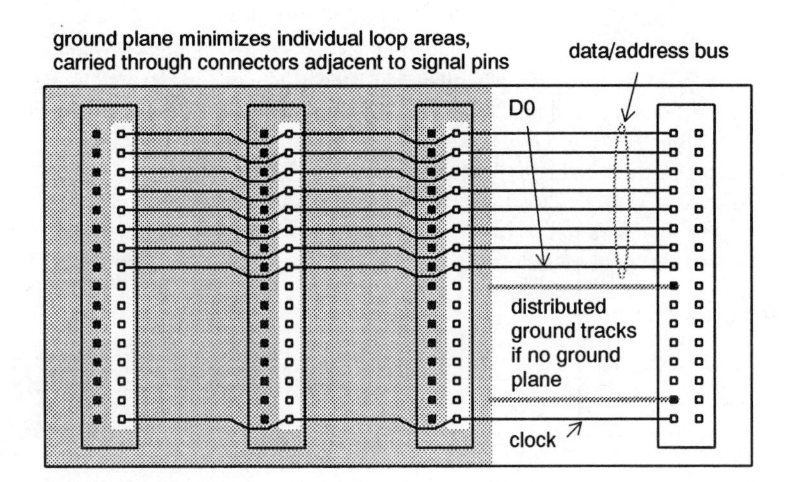

Figure 5.23 Backplane layout

an adjacent ground return.

5.2.2.5 *Ringing on transmission lines*

If you transmit data or clocks down long lines, these must be terminated to prevent ringing. Ringing is generated on the transitions of digital signals when a portion of the signal is reflected back down the line due to a mismatch between the line impedance and the terminating impedance. A similar mismatch at the driving end will re-reflect a further portion towards the receiver, and so on. Severe ringing will affect the data transfer if it exceeds the device's input noise margin.

Aside from its effect on noise margins, ringing may also be a source of interference in its own right. The amplitude of the ringing depends on the degree of mismatch at either end of the line while the frequency depends on the electrical length of the line (Figure 5.24). A digital driver/receiver combination should be analysed in terms of its transmission line behaviour if

$$2 \times t_{PD} \times \text{line length} \quad > \quad \text{transition time} \qquad (5.6)$$

where t_{PD} is the line propagation delay in ns per unit length [60]

Line propagation delay itself depends on the dielectric constant of the board material and can be calculated from section C.5 (Appendix C). This means matching the track's characteristic impedance to the source and load impedances, and may require extra components to terminate the line at the load. Most digital circuit data and application handbooks (e.g. [9],[13]) include advice and formulae for designing transmission line systems in fast logic. Table 5.3 is included as an aid to deciding whether the particular circuit you are concerned with should incorporate transmission line principles.

5.2.2.6 *Digital circuit decoupling*

No matter how good the V_{CC} and ground connections are, track distance will introduce an impedance which will create switching noise from the transient switching currents. The purpose of a decoupling capacitor is to maintain a low dynamic impedance from the individual IC supply voltage to ground. This minimises the local supply voltage

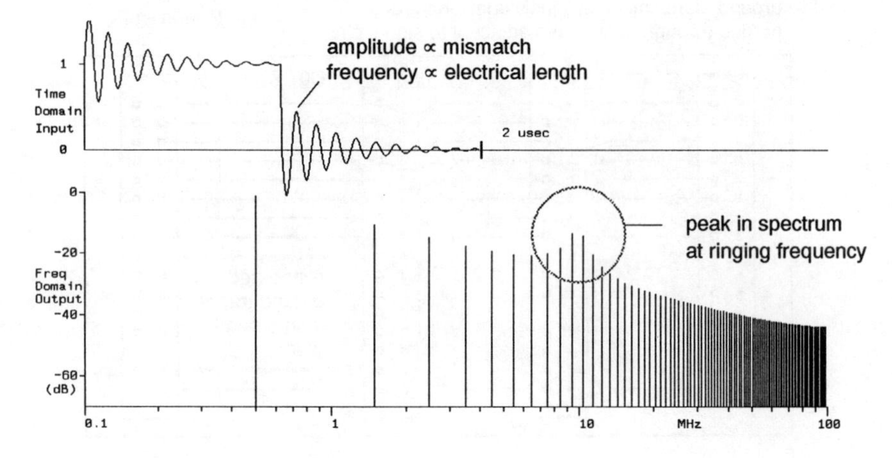

Figure 5.24 Ringing due to a mismatched transmission line

Logic family	t_r/t_f ns	Critical line length
4000B CMOS @ 5V	40	12 feet
74HC	6	1.75 feet
74LS	6	1.75 feet
74ALS	3.5	1 foot
74AC	3	10 inches
74F	3	10 inches
74AS	1.4	5 inches

Line length calculated for dielectric constant = 4.5 (FR4 epoxy glass), t_{PD} = 1.7ns/ft

Table 5.3 Critical transmission line length

droop when a fast current pulse is taken from it, and more importantly it minimizes the lengths of track which carry high di/dt currents. Placement is critical; the capacitor must be tracked close to the circuit it is decoupling. "Close" in this context means less than half an inch for fast logic such as AS-TTL, AC or ECL, especially when high current devices such as bus drivers are involved, extending to several inches for low-current, slow devices such as 4000B-series CMOS.

Components

The crucial factor when selecting capacitor type for high-speed logic decoupling is lead inductance rather than absolute value. Minimum lead inductance offers a low impedance to fast pulses. Small disk or multilayer ceramics, or polyester film types (lead pitch 2.5 or 5mm), are preferred; chip capacitors are even better. The overall inductance of each connection is the sum of both lead and track inductances. Flat ceramic capacitors, matched to the common dual-in-line pinouts and intended for mounting directly beneath the IC package, minimize the pin-to-pin inductance and offer superior performance above about 50MHz. They are appropriate for extending the usefulness of double sided boards (with a gridded ground layout but no ground plane) to clock frequencies approaching 50MHz. A recommended decoupling regime [18] for standard logic (74HC) is

- one 22μF bulk capacitor per board at the power supply input
- one 1μF tantalum capacitor per 10 packages of SSI/MSI logic or memory
- one 1μF tantalum capacitor per 2-3 LSI packages
- one 22nF ceramic or polyester capacitor for each octal bus buffer/driver IC or for each MSI/LSI package
- one 22nF ceramic or polyester capacitor per 4 packages of SSI logic

The value of 22nF offers a good tradeoff between medium frequency decoupling ability and high self resonant frequency (*cf* section 6.2.1.2). The minimum required value can be calculated from equation (5.7):

$$C = \Delta I \cdot \Delta t / \Delta V \quad (5.7)$$

ΔI and Δt can to a first order be taken from the figures in Table 5.1 and Table 5.2 while ΔV depends on your judgement of permissible supply voltage droop at the capacitor. Typically a power rail droop of 0.25V is reasonable; for an octal buffer taking 50mA per output and switching in 6ns, the required capacitance is 9.6nF. For smaller devices and faster switching times, less capacitance is required, and often the optimum capacitance value is as low as 1000pF. The lower the capacitance the higher will be its self resonant frequency, and the more effectively will it decouple the higher order harmonics of the switching current.

Small tantalum capacitors are to be preferred for bulk decoupling because due to their non-wound construction their self inductance is very much less than for an aluminium electrolytic of the same value.

5.2.3 Analogue circuit emissions

In general analogue circuits do not exhibit the high di/dt and fast risetimes that characterize digital circuits, and are therefore less responsible for excessive emissions. Analogue circuits which deliberately generate high frequency signals (remembering that the emissions regulatory regime currently begins at 150kHz, and may be extended downwards) need to follow the same layout and grounding rules as already outlined. It is also possible for low frequency analogue circuits to operate unintentionally outside their design bandwidth.

5.2.3.1 Instability

Analogue amplifier circuits may oscillate in the MHz region and thereby cause interference for a number of reasons:

- feedback-loop instability
- poor decoupling
- output stage instability

Capacitive coupling due to poor layout and common-impedance coupling are also sources of oscillation. Any prototype amplifier circuit should be checked for HF instability, whatever its nominal bandwidth, in its final configuration. Feedback instability is due to too much feedback near the unity-gain frequency, where the amplifier's phase margin is approaching a critical value. It may be linked with incorrect compensation of an uncompensated op-amp.

5.2.3.2 Decoupling

Power supply rejection ratio falls with increasing frequency, and power supply coupling to the input at high frequencies can be significant in wideband circuits. This is cured by decoupling, but typical 0.01 – 0.1μF decoupling capacitors may resonate with the parasitic inductance of long power leads in the MHz region, so decoupling-related instability problems usually show up in the 1 – 10MHz range. Paralleling a low-value capacitor with a 1 – 10μF tantalum capacitor will drop the resonant frequency and stray circuit Q to a manageable level. Note that the tantalum's series inductance could resonate with the ceramic capacitor and actually worsen the situation. To cure this, a few ohms resistance in series with the tantalum is necessary. The input stages of multi-stage high gain amplifiers may need additional resistance or a ferrite bead suppressor in series with each stage's supply to improve decoupling from the power rails.

5.2.3.3 Output stage instability

Capacitive loads cause a phase lag in the output voltage by acting in combination with the op-amp's open-loop output resistance (Figure 5.25) . This increased phase shift reduces the phase margin of a feedback circuit, possibly by enough to cause oscillation. A typical capacitive load, often invisible to the designer because it is not treated as a component, is a length of coaxial cable. Until the length starts to approach a quarter-wavelength at the frequency of interest, coax looks like a capacitor: for instance, 10 metres of the popular RG58C/U 50Ω type will be about 1000pF. To cure output instability, decouple the capacitance from the output with a low-value series resistor, and add high-frequency feedback with a small direct feedback capacitor C_F which compensates for the phase lag caused by C_L. When the critical frequency is high a ferrite bead is an acceptable substitute for R_S.

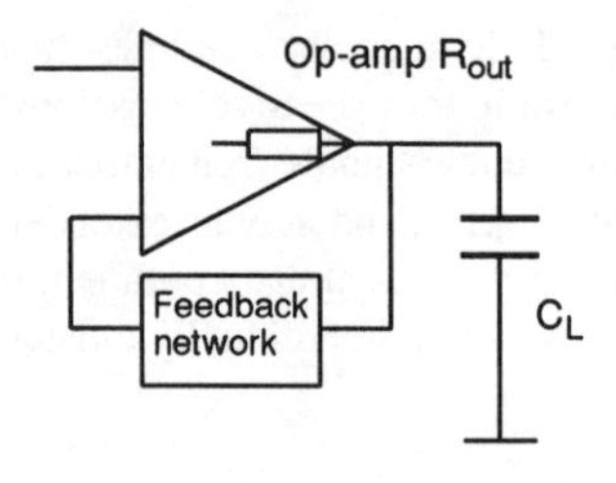

Phase lag @ freq f = $\tan^{-1}[f/f_c]$ degrees

where $f_c = 1 / 2\pi \cdot R_{out} \cdot C_L$

R_S = 10 – 100Ω

C_F

Feedback network

C_L

Isolate a large value of C_L with R_S and C_F, typically 20pF

Figure 5.25 Instability due to capacitive loads

5.2.4 The switching power supply

Switching supplies present extreme difficulties in containing generated interference [85]. Typical switching frequencies of 50 – 200kHz can be emitted by both differential and common mode mechanisms. Lower frequencies are more prone to differential mode emission while higher frequencies are worse in common mode. Figure 5.26 shows a typical direct-off-line switching supply with the major emission paths marked;

topologies may differ, or the transformer may be replaced by an inductor but the fundamental interference mechanisms are common to all designs.

5.2.4.1 Radiation from a high di/dt loop

Magnetic field radiation from a loop which is carrying a high di/dt can be minimized by reducing the loop area or by reducing di/dt. With low output voltages, the output rectifier and smoothing circuit may be a greater culprit in this respect than the input circuit. Loop area is a function of layout and physical component dimensions (see section 5.1.6). di/dt is a tradeoff against switching frequency and power losses in the switch. It can to some extent be controlled by slowing the rate-of-rise of the drive waveform to the switch.

The trend towards minimizing power losses and increasing frequencies goes directly against the requirements for low EMI. The lowest di/dt for a given frequency is given by a sinewave: sinusoidal converters (such as the series resonant converter, see [67]) have reduced EM emissions.

Magnetic component construction

Note that as explained in section 4.3.4.1 screening will have little effect on the magnetic field radiation due to this current loop although it will reduce the associated electric field. The transformer (or inductor) core should be in the form of a closed magnetic circuit in order to restrict magnetic radiation from this source. A toroid is the optimum from this point of view, but may not be practical because of winding difficulties or power losses; if you use a gapped core such as the popular E-core type, the gap should be directly underneath the windings since the greatest magnetic leakage flux is to be found around the gap.

Direct radiation from the circuit via this route has usually fallen well below radiated

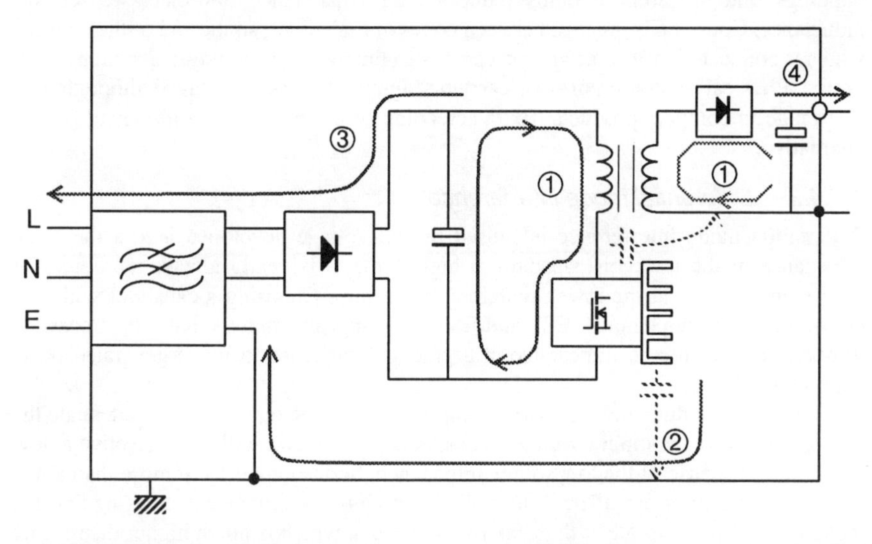

1: H-field radiation from high di/dt loop
2: capacitive coupling of E-field radiation from high dv/dt node to earth
3: differential-mode current conducted through DC link
4: conducted and/or radiated on output

Figure 5.26 Switching supply emission paths

emission limits at the lowest test frequency of 30MHz, unless the circuit rise times are very fast. On the other hand, such radiation can couple into the output or mains leads and be responsible for conducted interference (over 150kHz to 30MHz) if the overall power supply layout is poor. You should always keep any wiring which leaves the enclosure well away from the transformer or inductor.

5.2.4.2 Capacitive coupling to earth

High dv/dt at the switching point (the collector or drain of the switching transistor) will couple capacitively to ground and create common mode interference currents. The solution is to minimize dv/dt, and minimize coupling capacitance or provide a preferential route for the capacitive currents (Figure 5.27).

Dv/dt is reduced by a snubber and by keeping a low transformer leakage inductance and di/dt. These objectives are also desirable, if not essential, to minimize stress on the switching device, although they increase power losses. The snubber capacitor is calculated to allow a defined dv/dt with the maximum load as reflected through the transformer; the series resistor must be included to limit the discharge current through the switching device when it switches on. You can if necessary include a diode in parallel with the resistor to allow a higher resistor value and hence lower switching device ratings.

Capacitive screening

Capacitive coupling is reduced by providing appropriate electrostatic screens, particularly in the transformer and on the device heatsink. Note the proper connection of the screen: to either supply rail, which allows circulating currents to return to their source, not to earth. Even if the transformer is not screened, its construction can aid or hinder capacitive coupling from primary to secondary (Figure 5.27(b)). Separating the windings onto different bobbins reduces their capacitance but increases leakage inductance. Coupling is greatest between nodes of high dv/dt; so the end of the winding which is connected to V_{CC} or ground can screen the rest of the winding in a multi-layer design. Physical separation of parts carrying high dv/dt is desirable [85] although hard to arrange in compact products. Extra screening of the offending component(s) is an alternative.

5.2.4.3 Differential mode interference

Differential mode interference is caused by the voltage developed across the finite impedance of the reservoir capacitor at high di/dt. It is nearly always the dominant interference source at the lower switching harmonics. Choosing a capacitor with low equivalent series impedance (ESL and ESR) will improve matters, but it is impossible to obtain a low enough impedance in a practical capacitor to make generated noise negligible.

Extra series inductance and parallel capacitance on the input side will attenuate the voltage passed to the input terminals. A capacitor on its own will be ineffective at low frequencies because of the low source impedance. Series inductors of more than a few tens of microhenries are difficult to realize at high dc currents (remembering that the inductor must not saturate at the peak ripple current, which is much higher than the dc average current), and multiple sections with smaller inductors will be more effective than a single section. When several parallel reservoir capacitors are used, one of these may be separated from the others by the series inductor; this will have little effect on the overall reservoir but will offer a large attenuation to the higher frequency harmonics at little extra cost.

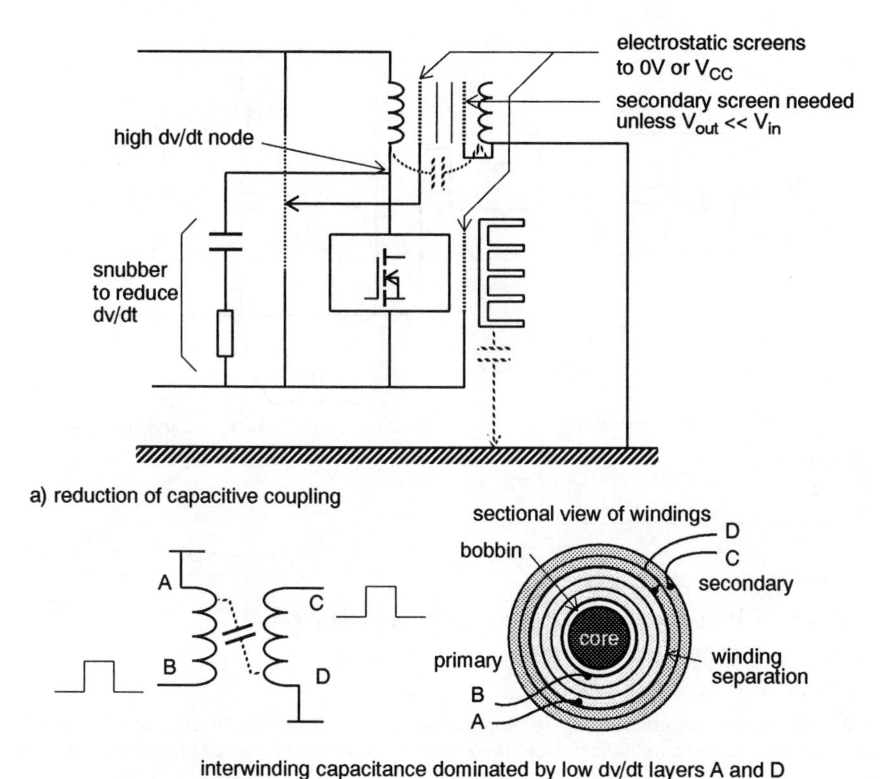

Figure 5.27 Common mode capacitive coupling

Figure 5.28 demonstrates filtering arrangements. The LC network may also be placed on the input side of the rectifier. This will have the advantage of attenuating broadband noise caused by the rectifier diodes switching at the line frequency. The mains input filter itself (see section 6.2.3) will offer some differential mode rejection. It is also possible to choose switching converter topologies with input inductors (such as the Cúk circuit [35]) which obviate fast di/dt transitions in the input and/or output waveforms. When testing the performance of a differential mode filter, be sure always to check it at the maximum operating input power. Not only do the higher switching currents generate more noise, but the peak mains input current may drive the filter inductor(s) into saturation and make it ineffective.

5.2.4.4 Output noise

Switching spikes are a feature of the dc output of all switching supplies, again mainly because of the finite impedance of the output reservoir. Such spikes are conducted out of the unit on the output lines in both differential and common mode, and may re-radiate onto other leads or be coupled to the ground connection and generate common mode interference. A low-ESL reservoir capacitor is preferable, but good differential mode suppression can be obtained, as with the input, with a high frequency L-section filter.

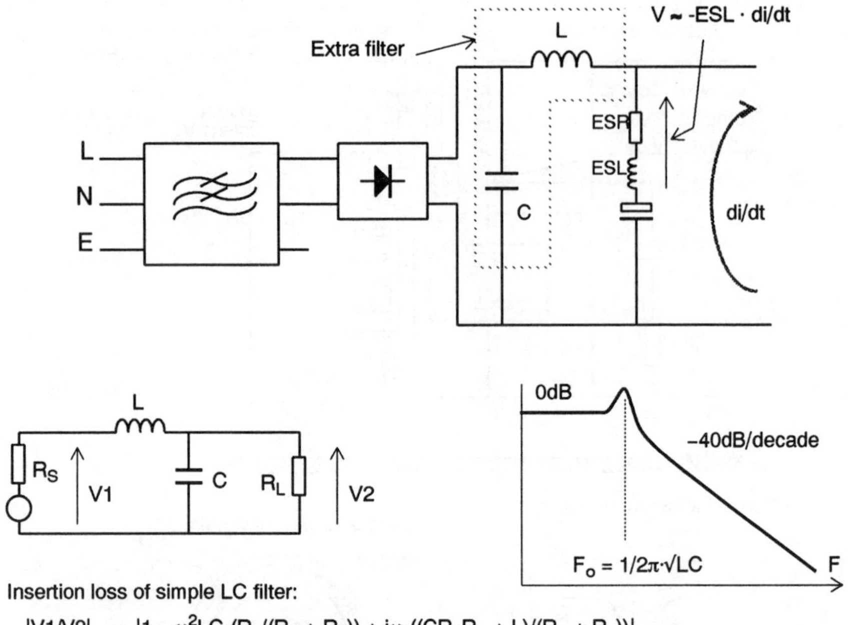

Insertion loss of simple LC filter:

$$|V1/V2| = |1 - \omega^2 LC \cdot (R_L/(R_S + R_L)) + j\omega \cdot ((CR_L R_S + L)/(R_S + R_L))|$$

For low R_S and high R_L this reduces to

$$|V1/V2| = |1 - \omega^2 LC|$$

Standard mains impedance R_L is approximated by 100Ω//100μH differentially, which approaches 100Ω above 300kHz. Note resonance of LC at low frequency gives insertion gain

Figure 5.28 Differential mode filtering

20–40dB is obtainable with a ferrite bead and 0.1μF capacitor above 1MHz. Common mode spikes will be unaffected by adding a filter, and this is a good way to diagnose which mode of interference is being generated.

The abrupt reverse recovery characteristic of the output rectifier diode(s) can create extra high frequency ringing and transients. These can be attenuated by using soft recovery diodes or by paralleling the diodes with an RC snubber.

5.2.5 Design for immunity – digital circuits

The first principle with microprocessor susceptibility is that because the logic threshold is nearer to 0V than to V_{CC}, most of the critical interference is ground-borne, whether it is common mode RF or transients. Differential mode interference will not propagate far into the circuit from the external interfaces. Therefore, lay out the circuit to keep ground interference currents away from the logic circuits. If layout is not enough, filter the I/O leads or isolate them, to define a preferential safe current path for interference. Radiated RF fields that generate differential mode voltages internally are dealt with in the same way as differential RF emissions, by minimizing circuit loop area, and by restricting the bandwidth of susceptible circuits where this is feasible.

Use the highest noise threshold logic family that is possible. 74HC is the best all-rounder. 4000B-series is useful for slow circuits, but beware of capacitively coupled interference. Use synchronous design, and try to avoid edge triggered data inputs wherever possible. However good the circuit's immunity, there will always be a transient that will defeat it. Every microprocessor should include a watchdog, and software techniques should be employed that minimize the effects of corruption.

Logic immunity principles

- Keep interference paths away from critical logic circuitry
  - layout
  - I/O filters and isolation
- Use high-noise-threshold logic
- Use a watchdog
- Adopt defensive programming tactics

5.2.5.1 Interference paths – transients

A tyypical microprocessor based product, including power supply, operator interface, processor board, enclosure and external connections can be represented at high frequency [21] by the layout shown in Figure 5.29. Note that the 0V rail will appear as a network of inductances with associated stray capacitances to the enclosure. An incoming common mode transient on the mains can travel through the circuit's 0V rail, generating ground differential spikes as it goes, through any or all of several paths as shown (observe the influential effect that stray capacitance has on these paths):

1: stray capacitance through the power supply to 0V, through the equipment and then to case

2: as above, but then out via an external connection

3: direct to case, then via stray capacitance to 0V and out via an external connection

If there are no external connections, (1) is the only problem and can be cured by a mains filter and/or by an electrostatic screen in the mains transformer. (2) arises because the external connection can provide a lower-impedance route to ground than case capacitance. You cannot control the impedance to ground of external connections, so you have to accept that this route will exist and ensure that the transient current has a preferential path via the case to the interface which does not take in the circuit. This is achieved by ensuring that the case structure is well bonded together, i.e. it presents a low impedance path to the transient, and by decoupling interfaces to the case at the point of entry/exit (see section 6.2.4 and section 5.1.7). If the enclosure is non-conductive then transient currents would have no choice but to flow through the circuit, and local grouping of interfaces is essential.

With external connections, route (3) can actually be *caused* by a mains filter, since at high frequencies parts of the enclosure can float with respect to true ground. This is sometimes the hardest concept to grasp – that even large conducting structures can exhibit high impedances, and hence voltage differentials when subjected to fast transients. The safety earth – the green and yellow wire – is *not* a reference point at HF (refer back to section 5.1.3.5 for ground wire impedances) and if this is the only "earth"

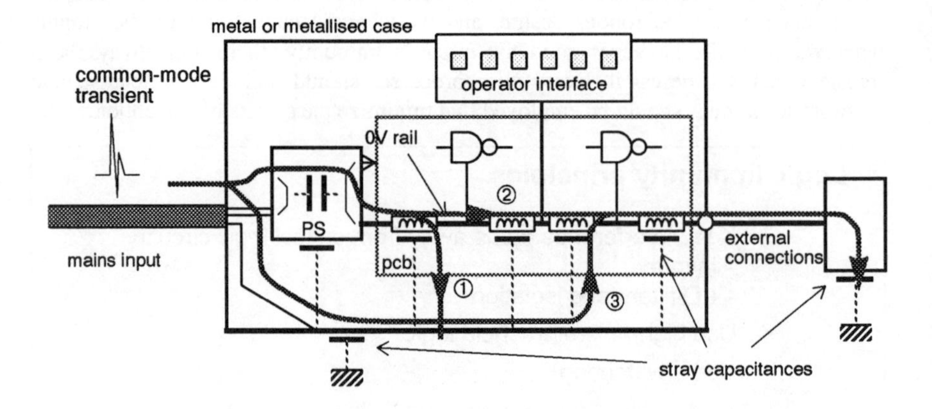

Figure 5.29 Representative high-frequency equivalent circuit: transients

connection, the cases's potential with respect to reference ground is defined by a complex network of inductances (connected cables) and stray capacitances which are impossible to predict. In all cases, grouping all I/O leads together with the mains lead (see Figure 5.16 on page 137) will offer low-inductance paths that bypass the circuit and prevent transient currents from flowing through the pcb tracks.

5.2.5.2 Interference paths – ESD

An electrostatic discharge can occur to any exposed part of the equipment. Common trouble spots as shown in Figure 5.30 are keyboards and controls (1), external cables (2) and accessible metalwork (3). A discharge to a nearby conductive object (which could be an ungrounded metal panel on the equipment itself) causes high local transient currents which will then also induce currents within the equipment by inductive or common impedance coupling.

Because there are many potential points of discharge, the possible routes to ground that the discharge current can take are widespread. Many of them will include part of the pcb ground layout, via stray capacitance, external equipment or exposed circuitry, and the induced transient ground differentials will cause maloperation. The discharge current will take the route (or routes) of least inductance. If the enclosure is well bonded to ground then this will be the natural sink point. If it is not, or if it is non-conductive, then the routes of least inductance will be via the connecting cables. If the edge of the pcb may be exposed, as in card frames, then a useful trick is to run a "guard trace" around it, unconnected to any circuitry, and separately bond this to ground.

When the enclosure consists of several conductive panels then these must all be well bonded together, following the rules described in section 6.3 for shielded enclosures. If this is not done then the edges of the panels will create very high transient fields as the discharge current attempts to cross them. If they are interconnected by lengths of wire, the current through the wire will cause a high magnetic field around it which will couple effectively with nearby pc tracks.

The discharge edge has an extremely fast risetime (sub-nanosecond, see section 4.3.3.1) and so stray capacitive coupling is essentially transparent to it, whilst even short ground connectors of a few nH will present a high impedance. For this reason the

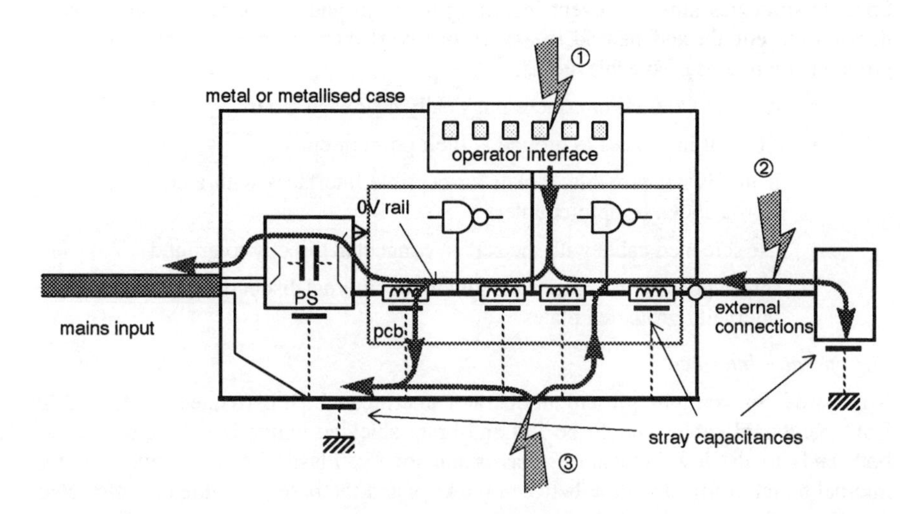

Figure 5.30 Representative high-frequency equivalent circuit: ESD

presence or absence of a safety ground wire (which has a high inductance) will make little difference to the system response to ESD.

5.2.5.3 Transient and ESD protection

Techniques to guard against corruption by transients and ESD are generally similar to those used to prevent RF emissions, and the same components will serve both purposes.

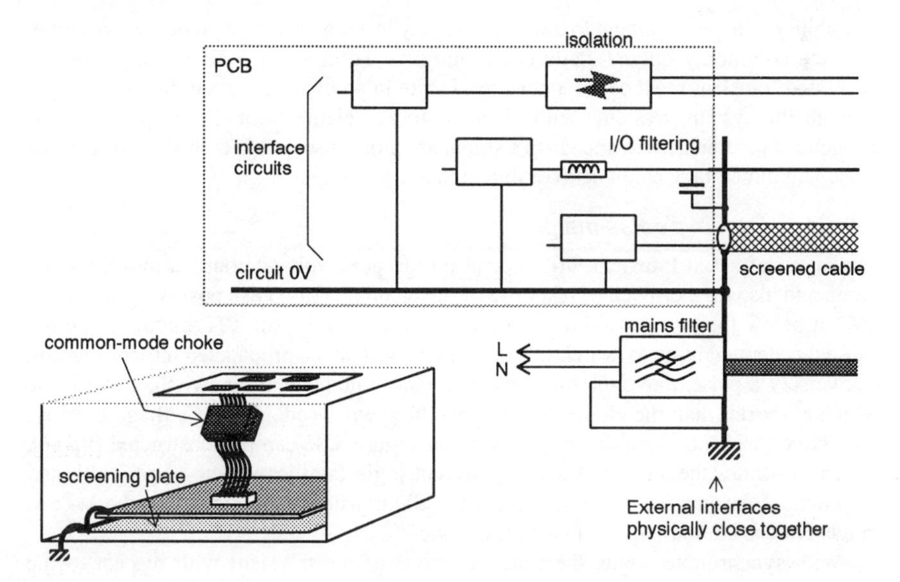

Figure 5.31 ESD protection

Specific strategies aim to prevent incoming transient and RF currents from flowing through the circuit, and instead to absorb or divert them harmlessly and directly to ground (Figure 5.31). To achieve this,

- keep all external interfaces physically near each other
- filter all interfaces to ground at their point of entry
- if this is not possible, isolate susceptible interfaces with a common-mode ferrite choke or opto-couplers
- use screened cable with the screen connected directly to ground
- screen pcbs from exposed metalwork or external discharge points with extra internally grounded plates

The operator interface

Keyboards, for example, present an operator interface which is frequently exposed to ESD. Keyboard cables should be foil-and-braid shielded which is 360° grounded at both ends to the low-inductance chassis metalwork. Plastic key caps will call for internal metal or foil shielding between the keys and the base pcb which is connected directly to the cable shield, to divert transients away from the circuitry. The shield ground should be coupled to the circuit ground at the cable entry point via a 10 – 100nF capacitor to prevent ground potential separation during an ESD event. A membrane keyboard with a polyester surface material has an inherently high dielectric strength and is therefore resistant to ESD, but it should incorporate a ground plane to provide a bleed path for the accumulated charge and to improve RF immunity; this ground plane must be "hidden" from possible discharges by sealing it behind the membrane surface.

5.2.6 Logic noise immunity

The ability of a logic element to operate correctly in a noisy environment involves more than the commonly quoted static noise margins. To create a problem an externally generated transient must cause a change of state in an element which then propagates through the system. Systems with clocked storage elements or those operating fast enough for the transient to appear as a signal are more susceptible than slow systems or those without storage elements (combinational logic only).

5.2.6.1 Dynamic noise margin

The effect of a fast transient will depend on the peak voltage coupled into the logic input, and also on the speed of response of the element. Any pulse positive-going from 0V but below the logic switching threshold (typically 1.8V for TTL circuits) will not cause the element input to switch from 0 to 1 and will not be propagated into the system. Conversely a pulse above the threshold will cause the element to switch. But a pulse which is shorter than the element's response time will need a higher voltage to cause switchover, and therefore the graph shown in Figure 5.32 can be constructed [9][13], which illustrates the susceptibility of different logic families versus pulse width and amplitude. Bear in mind that switching and ESD transients may lie within the 1 – 5ns range. Here is another argument for slow logic!

With synchronous logic, the time of arrival of the transient with respect to the system clock (assuming it corrupts the data line rather than the clock line, due to the former's usually greater area) is important. If the transient does not coincide with the active clock edge then an incorrect value on the data line will not propagate through the system. Thus you can expand the graphs of Figure 5.32 to incorporate another

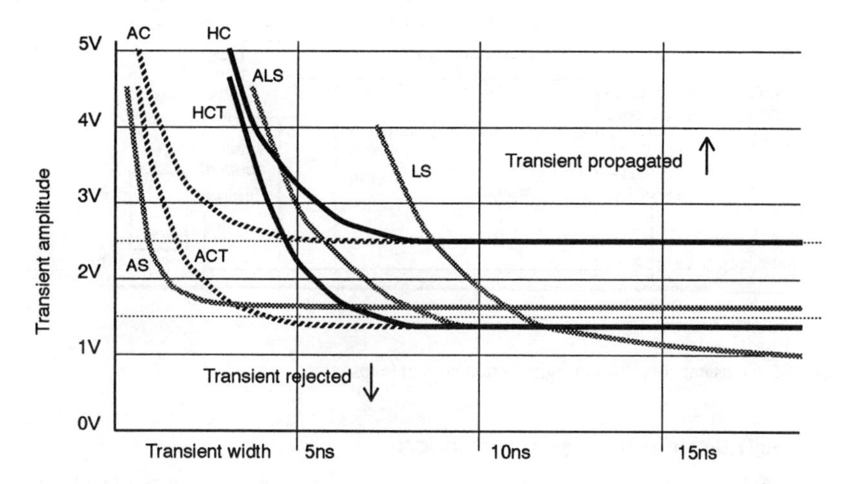

Figure 5.32 Dynamic noise margins

dimension of elapsed time from the clock edge. Tront [81] has simulated a combinational logic circuit with a flip-flop in 3-micron CMOS technology and generated a series of "upset windows" in this way to describe the susceptibility of that particular circuit to interference. Such a simulation process, using the simulation package SPICE3, can pinpoint those parts of a circuit which have a high degree of susceptibility.

5.2.6.2 Transient coupling

The amplitude of any pulse coupled differentially into a logic input will depend on the loop area of the differential coupling path which is subjected to the transient field $H_{transient}$ due to transient ground currents $I_{transient}$, and also on the impedance of the driving circuit – less voltage is coupled into a lower impedance. For this reason the lower logic 0 threshold voltage of LSTTL versus HCMOS is somewhat compensated by the higher logic 0 output sink capability of LSTTL. If sensitive signal tracks are run close to their ground returns as recommended for emissions, then the resulting loop area is small and little interference is coupled differentially into the sensitive input (Figure 5.33).

Susceptibility prediction

IC susceptibility to RF, as opposed to transient interference, tends to be most marked in the 20 – 200MHz region. Susceptibility at the component level is broadband, although there are normally peaks at various frequencies due to resonances in the coupling path. As the frequency increases into the microwave region component response drops as parasitic capacitances offer alternative shunt paths for the RF energy, and the coupling becomes less efficient. Prediction of the level of RF susceptibility of digital circuits using simulation is possible for small scale integrated circuits [84] but the modelling of the RF circuit parameters of VLSI devices requires considerable effort, and the resources needed to develop such models for microprocessors and their associated peripherals is overwhelmed by the rate of introduction of new devices.

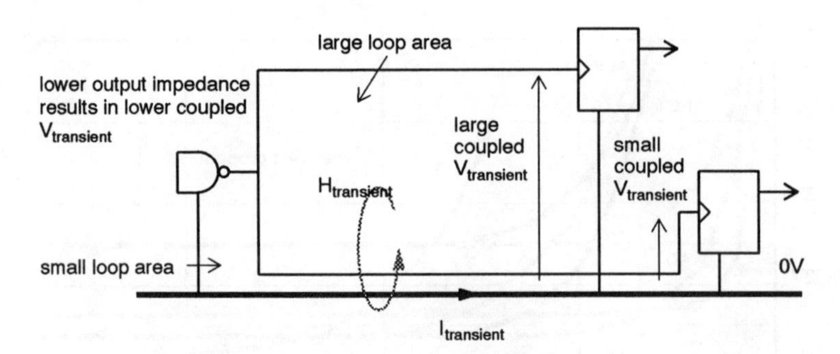

Figure 5.33 Transient coupling via signal/return current loops

5.2.7 The microprocessor watchdog

Circuit techniques to minimise the amplitude and control the path of disruptive interference go a long way towards "hardening" a microprocessor circuit against corruption. But they cannot *eliminate* the risk. The coincidence of a sufficiently high-amplitude transient with a vulnerable point in the data transfer is an entirely statistical affair. The most cost-effective way to ensure the reliability of a microprocessor-based product is to accept that the program *will* occasionally be corrupted, and to provide a means whereby the program flow can be automatically recovered, preferably transparently to the user. This is the function of the microprocessor watchdog [18].

Some of the more up-to-date micros on the market include built-in watchdog devices, which may take the form of an illegal-opcode trap, or a timer which is repetitively reset by accessing a specific register address. If such a watchdog is available, it should be used, because it will be well matched to the processor's operation; otherwise, one must be designed-in to the circuit.

5.2.7.1 Basic operation

The most serious result of a transient corruption is that the processor program counter or address register is upset, so that it starts interpreting data or empty memory as valid instructions. This causes the processor to enter an endless loop, either doing nothing or performing a few meaningless or, in the worst case, dangerous instructions. A similar effect can happen if the stack register or memory is corrupted. Either way, the processor will appear to be catatonic, in a state of "dynamic halt".

A watchdog guards against this eventuality by requiring the processor to execute a specific simple operation regularly, regardless of what else it is doing, on pain of consequent reset. The watchdog is actually a timer whose output is linked to the $\overline{\text{RESET}}$ input, and which itself is being constantly retriggered by the operation the processor performs, normally writing to a spare output port. This operation is shown schematically in Figure 5.34.

5.2.7.2 Timeout period

If the timer does not receive a "kick" from the output port for more than its timeout period, its output goes low ("barks") and forces the microprocessor into reset. The timeout period must be long enough so that the processor does not have to interrupt time-critical tasks to service the watchdog, and so that there is time for the processor to

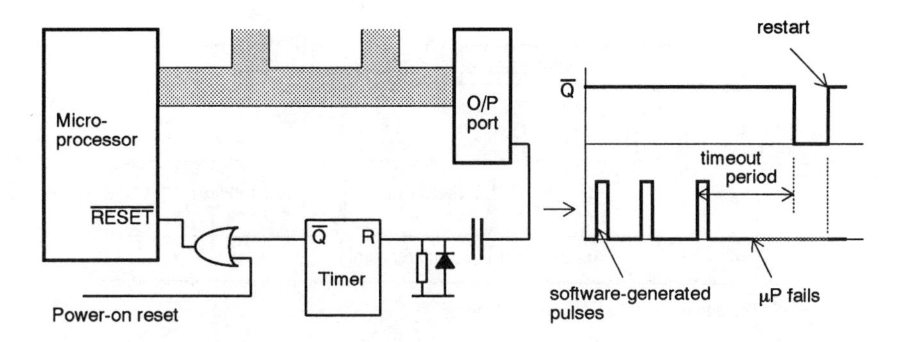

Figure 5.34 Watchdog operation

start the servicing routine when it comes out of reset (otherwise it would be continually barking and the system would never restart properly). On the other hand, it must not be so long that the operation of the equipment could be corrupted for a dangerous period. There is no one timeout period which is right for all applications, but usually it is somewhere between 10ms and 1s.

5.2.7.3 Timer hardware

The watchdog circuit has to exceed the reliability of the rest of the circuit and so the simpler it is, the better. A standard timer IC is quite adequate, but the timeout period may have an unacceptably wide variation in tolerance, besides needing extra discrete components. A digital divider such as the 4060B fed from a high-frequency clock and periodically reset by the report pulses is a more attractive option, since no other components are needed. The divider logic could instead be incorporated into an ASIC if this is present for other purposes. The clock has to have an assured reliability in the presence of transient interference, but such a clock may well already be present or could be derived from the unsmoothed AC input at 50/60Hz.

An extra advantage of the digital divider approach is that its output in the absence of retriggering is a stream of pulses rather than a one-shot. Thus if the micro fails to be reset after the first pulse, or more probably is derailed by another burst of interference before it can retrigger the watchdog, the watchdog will continue to bark until it achieves success (Figure 5.35). This is far more reliable than a monostable watchdog that only barks once and then shuts up.

A programmable timer must not be used to fulfil the watchdog function, however attractive it may be in terms of component count. It is quite possible that the transient corruption could result in the timer being programmed off, thereby completely silencing the watchdog.Similarly, it is unsafe to disable the watchdog from the program while performing long operations; corruption during this period will not be recoverable. It is better to insert extra watchdog “kicks” during such long sequences.

5.2.7.4 Connection to the microprocessor

Figure 5.34 shows the watchdog’s $\overline{Q}$ output being fed directly to the $\overline{RESET}$ input along with the power-on reset (POR) signal. In many cases it will be possible and preferable to trigger the timer’s output from the POR signal, in order to assure a defined reset pulse width at the micro on power-up.

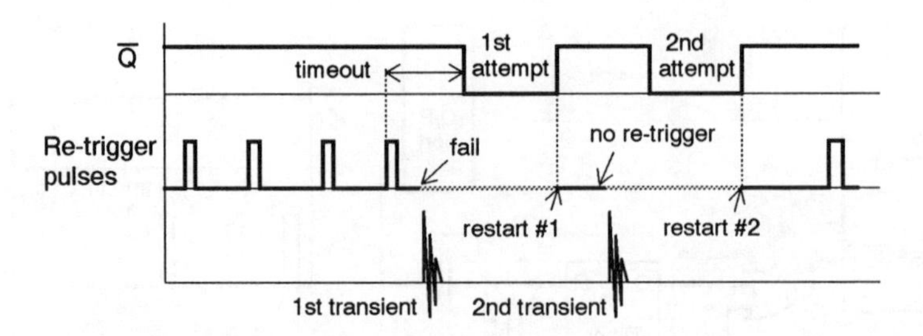

Figure 5.35 The advantage of an astable watchdog

It is essential to use the RESET input and not some other signal to the micro such as an interrupt, even a non-maskable one. The processor may be in any conceivable state when the watchdog barks, and it must be returned to a fully characterised state. The only state which can guarantee a proper restart is RESET. If the software must know that it was the watchdog that was responsible for the reset, this should be achieved by reading a separate latched input port during initialization.

5.2.7.5 Source of the re-trigger pulse

Equally important is that the micro should not be able to carry on kicking the watchdog when it is catatonic. This demands AC coupling to the timer's re-trigger input, as shown by the R-C-D network in Figure 5.34. This ensures that only an edge will re-trigger the watchdog, and prevents an output which is stuck high or low from holding the timer off. The same effect is achieved with a timer whose re-trigger input is edge- rather than level-sensitive.

Using a programmable port output in conjunction with AC coupling is attractive for two reasons. It needs two separate instructions to set and clear it, making it very much less likely to be toggled by the processor executing an endless loop; this is in contrast to designs which use an address decoder to produce a pulse whenever a given address is accessed, which practice is susceptible to the processor rampaging uncontrolled through the address space. Secondly, if the programmable port device is itself corrupted but processor operation otherwise continues properly, then the retrigger pulses may cease even though the processor is attempting to write to the port. The ensuing reset will ensure that the port is fully re-initialised. As a matter of software policy, programmable peripheral devices should be periodically re-initialised anyway.

5.2.7.6 Generation of the re-trigger pulses in software

If possible, two independent software modules should be used to generate the two edges of the report pulse (Figure 5.36). With a port output as described above, both edges are necessary to keep the watchdog held off. This minimises the chance of a rogue software loop generating a valid re-trigger pulse. At least one edge should only be generated at one place in the code; if a real-time "tick" interrupt is used, this could be conveniently placed at the entry to the interrupt service routine, whilst the other is placed in the background service module. This has the added advantage of guarding against the interrupt being accidentally masked off.

Placing the watchdog re-trigger pulse(s) in software is the most critical part of watchdog design and repays careful analysis. On the one hand, too many calls in different modules to the pulse generating routine will degrade the security and

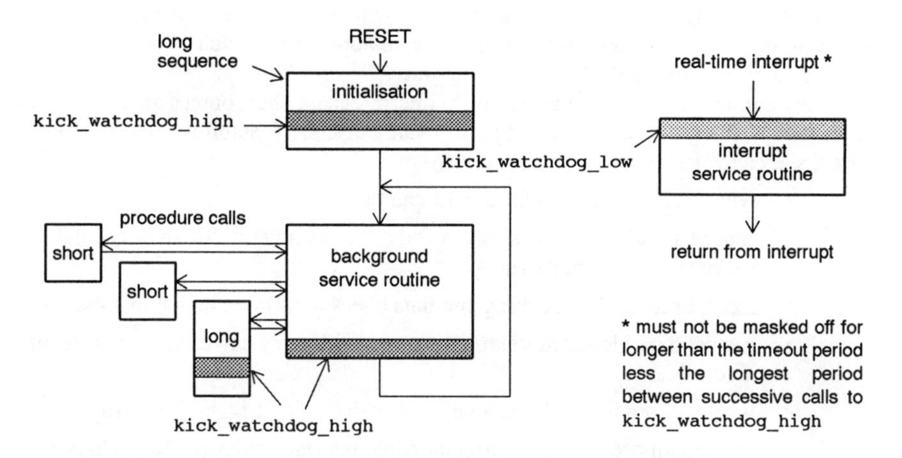

Figure 5.36 Software routine for watchdog re-trigger

efficiency of the watchdog; but on the other hand, any non-trivial application software will have execution times that vary and will use different modules at different times, so that pulses will have to be generated from several different places. Two frequent critical points are on initialisation, and when writing to non-volatile (EEPROM) memory. These processes may take several tens of milliseconds. Analysing the optimum placement of re-trigger pulses, and ensuring that under all correct operating conditions they are generated within the timeout period, is not a small task.

5.2.7.7 Testing the watchdog

This is not at all simple, since the whole of the rest of the circuit design is bent towards making sure the watchdog never barks. Creating artificial conditions in the software is unsatisfactory because the tested system is then unrepresentative. An adequate procedure for most purposes is to subject the equipment to repeated transient pulses which are of a sufficient level to corrupt the processor's operation predictably, if necessary using specially "weakened" hardware. For safety critical systems you may have to perform a statistical analysis to determine the pulse repetition rate and duration of test that will establish acceptable performance. An LED on the watchdog output is useful to detect its barks. A particularly vulnerable condition is the application of a burst of spikes, so that the processor is hit again just as it is recovering from the last one. This is unhappily a common occurrence in practice.

As well as testing the reliability of the watchdog, a link to disable it must be included in order to test new versions of software.

5.2.8 Defensive programming

Some precautions against interference can be taken in software. Standard techniques of data validation and error correction should be widely used. Hardware performance can also be improved by well-thought-out software. Some means of disabling software error-checking is useful when optimizing the equipment hardware against interference, as otherwise weak points in the hardware will be masked by the software's recovery capabilities. For example, software which does not recognise digital inputs until three polls have given the same result will be impervious to transients which are shorter than this. If your testing uses only short bursts or single transients the equipment will appear to be immune, but longer bursts will cause maloperation which might have been prevented by improving the hardware immunity.

Not all microprocessor faults are due to interference. Other sources are intermittent connections, marginal hardware design, software bugs, meta-stability of asynchronous circuits, etc. Typical software techniques are to

- type-check and range-check all input data
- sample input data several times and either average it, for analogue data, or validate it, for digital data
- incorporate parity checking and data checksums in all data transmission
- protect data blocks in volatile memory with error detecting and correcting algorithms
- wherever possible rely on level- rather than edge-triggered interrupts
- periodically re-initialize programmable interface chips (PIAs, ACIAs, etc)

5.2.8.1 Input data validation and averaging

If you can set known limits on the figures that enter as digital input to the software then you can reject data which are outside those limits. When, as in most control or monitoring applications, each sensor inputs a continuous stream of data, this is simply a question of taking no action on false data. Since the most likely reason for false data is corruption by a noise burst or transient, subsequent data in the stream will probably be correct and nothing is lost by ignoring the bad item. Data-logging applications might require a flag on the bad data rather than merely to ignore it.

This technique can be extended if there is a known limit to the maximum rate-of-change of the data. An input which exceeds this limit can be ignored even though it may be still within the range limits. Software averaging on a stream of data to smooth out process noise fluctuations can also help remove or mitigate the effect of invalid data.

You should take care when using sophisticated software for error detection not to lock out genuine errors which need flagging or corrective action, such as a sensor failure. The more complex the software algorithm is, the more it needs to be tested to ensure that these abnormal conditions are properly handled.

5.2.8.2 Digital inputs

A similar checking process should be applied to digital inputs. In this case, there are only two states to check so range testing is inappropriate. Instead, given that the input ports are being polled at a sufficiently high rate, compare successive input values with each other and take no action until two or three consecutive values agree. This way, the processor will be "blind" to occasional transients which may coincide with the polling time slot. This does mean that the polling rate must be two or three times faster than the minimum required for the specified response time, which in turn may require a faster microprocessor than originally envisaged.

5.2.8.3 Interrupts

For similar reasons to those outlined above, it is preferable not to rely on edge-sensitive interrupt inputs. Such an interrupt can be set by a noise spike as readily as by its proper signal. Undoubtedly edge-sensitive interrupts are necessary in some applications, but in these cases you should treat them in the same way as clock inputs to latches or flip-flops and take extra precautions in layout and drive impedance to minimise their noise susceptibility. If there is a choice in the design implementation, then favour a level-sensitive interrupt input.

5.2.8.4 Data and memory protection

Volatile memory (RAM, as distinct from ROM or EEPROM) is susceptible to various forms of data corruption. These can be prevented by placing critical data in tables in RAM. Each table is then protected by a checksum, which is stored with the table. Checksum-checking diagnostics can be run by the background routine automatically at whatever interval is deemed necessary to catch RAM corruption, and an error can be flagged or a software reset can be generated as required. The absolute values of RAM data do not need to be known provided that the checksum is recalculated every time a table is modified. Beware that the diagnostic routine is not interrupted by a genuine table modification or vice versa, or errors will start appearing from nowhere! Of course, the actual partitioning of data into tables is a critical system design decision, as it will affect the overall robustness of the system.

5.2.8.5 Unused program memory

One of the threats discussed in the section on watchdogs (section 5.2.7) was the possibility of the microprocessor accessing unused memory space due to corruption of its program counter. If it does this, it will interpret whatever data it finds as a program instruction. In such circumstances it would be useful if this action had a predictable outcome.

Normally a bus access to a non-existent address returns the data $\#FF_H$, provided there is a passive pull-up on the bus, as is normal practice. Nothing can be done about this. However, un-programmed ROM also returns $\#FF_H$ and this can be changed. A good approach is to convert all unused $\#FF_H$ locations to the processor's one-byte NOP (no operation) instruction (Figure 5.37). The last few locations in ROM can be programmed with a JMP RESET instruction, normally three bytes, which will have the effect of resetting the processor. Then, if the processor is corrupted and accesses anywhere in unused memory, it finds a string of NOP instructions and executes these (safely) until it reaches the JMP RESET, at which point it restarts.

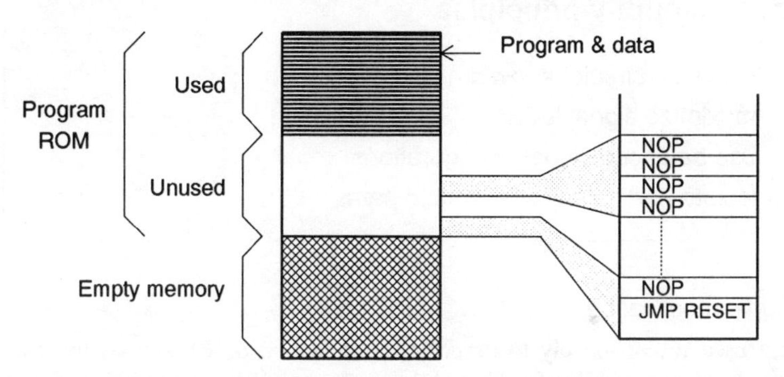

Figure 5.37 Protecting unused program memory with NOPs

The effectiveness of this technique depends on how much of the total possible memory space is filled with NOPs, since the processor can be corrupted to a random address. If the processor accesses an empty bus, its action will depend on the meaning of the $\#FF_H$

instruction. The relative cheapness of large ROMs and EPROMs means that you could consider using these, and filling the entire memory map with ROM, even if your program requirements are small.

5.2.8.6 Re-initialization

As well as RAM data, you must remember to guard against corruption of the set-up conditions of programmable devices such as I/O ports or UARTs. Many programmers assume erroneously that once an internal device control register has been set up (usually in the initialization routine) it will stay that way forever. Experience shows that control registers can change their contents, even though they are not directly connected to an external bus, as a result of interference. This may have consequences that are not obvious to the processor: for instance if an output port is re-programmed as an input, the processor will happily continue writing data to it oblivious of its ineffectiveness.

The safest course is to periodically re-initialize all critical registers, perhaps in the main idling routine if one exists. Timers, of course, cannot be protected in this way. The period between successive re-initializations depends on how long the software can tolerate a corrupt register, versus the software overhead associated with the re-initialization.

5.2.9 Transient and RF immunity - analogue circuits

Analogue circuits in general are not as susceptible to transient upset as digital, but may be more susceptible to demodulation of RF energy. This can show itself as a dc bias shift which results in measurement non-linearities or non-operation, or as detection of modulation, which is particularly noticeable in audio and video circuits. Such bias shift does not affect digital circuit operation until the bias is enough to corrupt logic levels, at which point operation ceases completely. Improvements in immunity result from attention to the four areas as set out below. The greatest rf signal levels are those coupled in via external interface cables and so interface circuits should receive the first attention.

Analogue immunity principles

- minimize circuit bandwidth
- maximize signal levels
- use balanced signal configurations
- isolate particularly susceptible paths

5.2.9.1 Audio rectification

This is a term used rather loosely to describe the detection of RF signals by low-frequency circuits. It is responsible for most of the ill effects of RF susceptibility of both analogue and digital products.

When a circuit is fed an RF signal that is well outside its normal bandwidth, the circuit can respond either linearly or non-linearly (Figure 5.38). If the signal level is low enough for it to stay linear, it will pass from input to output without affecting the wanted signals or the circuit's operation. If the level drives the circuit into non-linearity, then

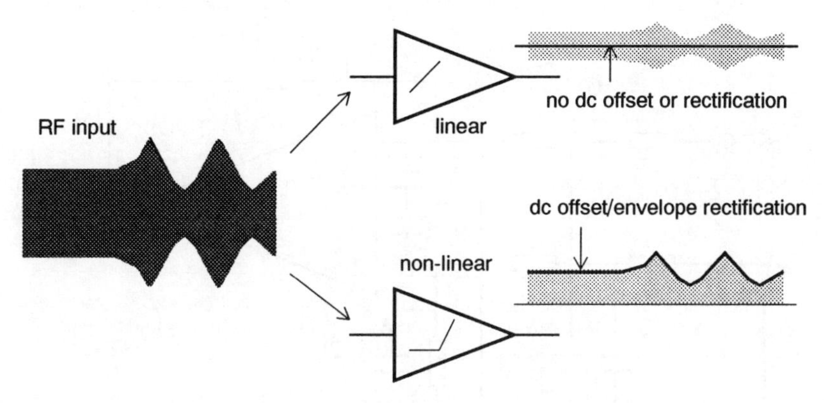

Figure 5.38 RF demodulation by non-linear circuits

the envelope of the signal (perhaps severely distorted) will appear on the circuit's output. At this point it will be inseparable from the wanted signal and indeed the wanted signal will itself be affected by the circuit's forced non-linearity. The response of the circuit depends on its linear dynamic range and on the level of the interfering signal. All other factors being equal, a circuit which has a wide dynamic range will be more immune to RF than one which has not.

5.2.9.2 Bandwidth, level and balance

The level of the interfering signal can be reduced by restricting the operating bandwidth to the minimum acceptable. This can be achieved (referring to Figure 5.39) by input RC or LC filtering (1), feedback RC filtering (2), and low value (10 – 33pF) capacitors directly at the input terminals (3). RC filters may degrade stability or worsen the circuit's common mode rejection (CMR) properties, and the value of C must be kept low to avoid this, but an improvement in RF rejection of between 10 to 35dB over the range 0.15 to 150MHz has been reported [79] by including a 27pF feedback capacitor on an ordinary inverting op-amp circuit. High frequency CMR is determined by the unbalance between capacitances on balanced inputs. If R would consequently be too high and might affect circuit dc conditions, a lossy ferrite-cored choke or bead is an alternative series element.

You should design for signal level to be as high as possible throughout, consistent with other circuit constraints, but at the same time impedances should also be maintained as low as possible to minimize capacitive coupling and these requirements may conflict. The decision will be influenced by whether inductive coupling is expected to be a major interference contributor. If it is (because circuit loop areas cannot be made acceptably small), then higher impedances will result in lower coupled interference levels. Refer to the discussion on inductive and capacitive coupling (section 4.1.1).

Balanced circuit configurations allow maximum advantage to be taken of the inherent common-mode rejection of op-amp circuits. But note that CMR is poorer at high frequencies and is affected by capacitive and layout imbalances, so it is unwise to trust too much in balanced circuits for good RF and transient immunity. It has also been observed [84] that in the frequency range 1 – 20MHz, mean values of demodulated RFI are 10 to 20dB lower for BiFET op amps than for bipolar op amps.

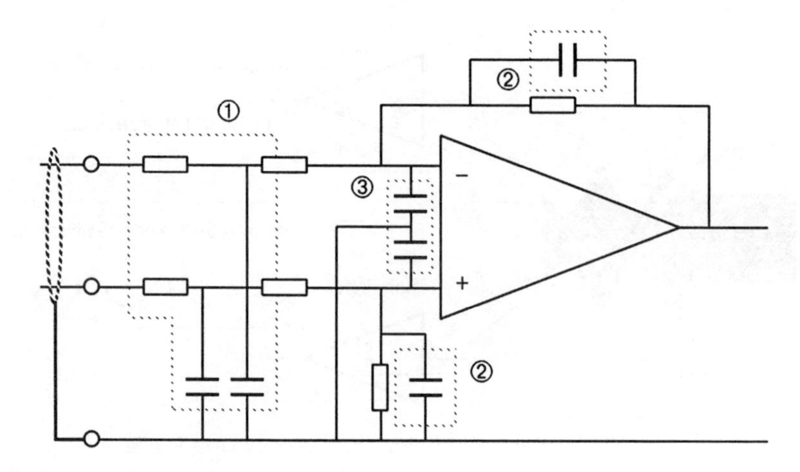

Figure 5.39 Bandwidth limitation

5.2.9.3 Isolation

Signals may be isolated at input or output with either an opto-coupler or a transformer (Figure 5.40). The ultimate expression of the former is fibre optic data transmission, which with the falling costs of fibre optic components is becoming steadily more attractive in a wide range of applications. Given that the major interference coupling route is via the connected cables, using optical fibre instead of wire completely removes this route. This leaves only direct coupling to the enclosure, and coupling via the power cable, each of which is easier to deal with than electrical signal interfaces. Signal processing techniques will be needed to ensure accurate transmission of precision ac or dc signals, which increases the overall cost and board area.

Coupling capacitance

Isolation breaks the electrical ground connection and therefore substantially removes common mode noise injection, as well as allowing a DC or low frequency AC potential difference to exist. However there is still a residual coupling capacitance which will compromise the isolation at high frequencies or high rates of common mode dv/dt. This capacitance is typically 3pF per device for an opto-coupler; where several channels are isolated the overall coupling capacitance (from one ground to the other) rises to several tens of pF. This common mode impedance is a few tens of ohms at 100MHz, which is not much of a barrier!

Electrostatically screened transformers and opto-couplers are available where the screen reduces the coupling of common mode signals into the receiving circuit, and hence improves the common mode transient immunity *of that (local) circuit.* This improvement is gained at the expense of increasing the overall capacitance across the isolation barrier and hence reducing the impedance of the transient or RF coupling path to the rest of the unit. A somewhat expensive solution to this problem is to use two unscreened transformers in series, with the intervening coupled circuit separately grounded.

It is best to minimize the number of channels by using serial rather than parallel data transmission. Do not compromise the isolation further by running tracks from one

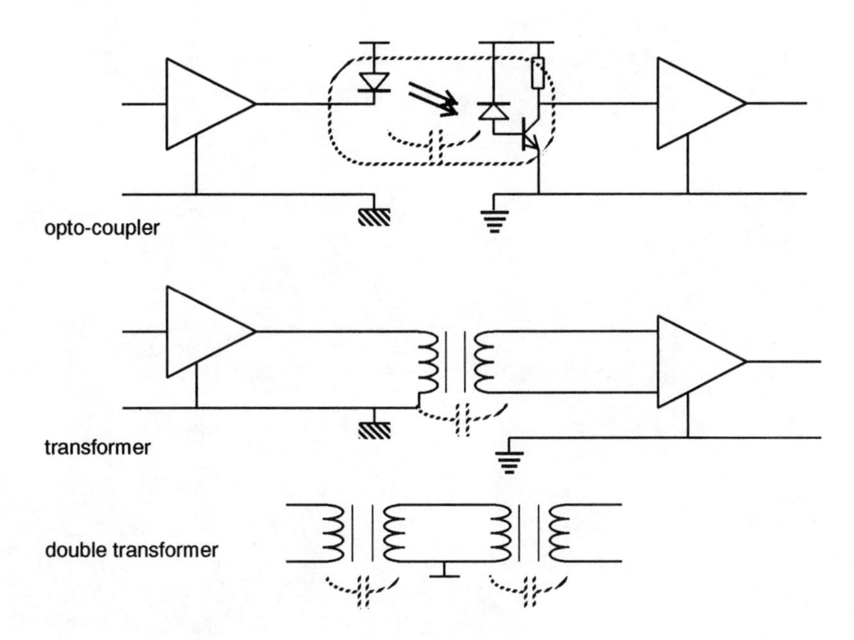

Figure 5.40 Signal isolation

circuit near to tracks from the other. Figure 4.4 on page 99 will give some idea of the degree of mutual coupling versus track spacing.

Chapter 6

Interfaces, filtering and shielding

6.1 Cables and connectors

The most important sources of radiation from a system, or of coupling into a system, are the external cables. Due to their length these are more efficient at interacting with the electromagnetic environment than enclosures, pcbs or other mechanical structures. Cables, and the connectors which form the interface to the equipment, must be carefully specified. The main purpose of this is to ensure that differential-mode signals are prevented from radiating from the cables, and that common-mode cable currents are neither impressed on the cable by the signal circuit nor are coupled into the signal circuit from external fields via the cable.

In many cases you will have to use screened cables. Exceptions are the mains power cable (provided a mains filter is fitted), and low-frequency interfaces which can be properly filtered to provide transient and RF immunity. An unfiltered, unscreened interface will provide a path for external emissions and for undesired inward coupling. The way that the cable screen is terminated at the connector interface is critical in maintaining the screening properties of the cable.

6.1.1 Cable segregation and returns

To minimize crosstalk effects within a cable, the signals carried by that cable should all be approximately equal (within, say, ±10dB) in current and voltage. This leads to the grouping of cable classifications shown in Figure 6.1. Cables carrying high frequency interfering currents should be kept away from other cables, even within shielded

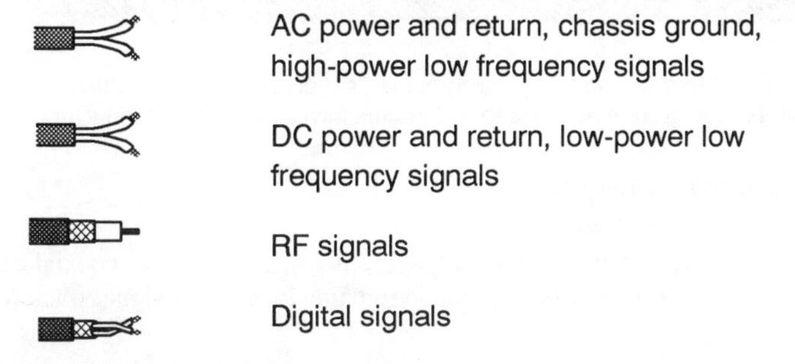

Figure 6.1 Cable classification

enclosures, as the interference can readily couple to others nearby and generate conducted common-mode emissions. See Figure 4.4 on page 99 for the effect on mutual capacitance and inductance of the spacing between cables. The breakpoint at which "low frequency" becomes "high frequency" is determined by cable capacitance and circuit impedances and may be as low as a few kHz.

All returns should be closely coupled to their signal or power lines, preferably by twisting, as this reduces magnetic field coupling to the circuit. Returns should never be shared between power and signal lines, and preferably not between individual signal lines, as this leads to common impedance coupling. Extra uncommitted grounded wires can help to reduce capacitive crosstalk within cables.

Having advised segregation of different cable classes, it is still true that the best equipment design will be one which puts no restrictions on cable routing and mixing – i.e. one where the major EMC design measures are taken internally. There are many application circumstances when the installation is carried out by unskilled and untrained technicians who ignore your carefully specified guidelines, and the best product is one which works even under these adverse circumstances.

Return currents

It is not intuitively obvious that return currents will necessarily flow in the conductor which is local to the signal wires, when there are several alternative return paths for them to take. At DC, the return currents are indeed shared only by the ratio of conductor resistances. But as the frequency increases the mutual inductance of the coupled pair (twisted or coaxial) tends to reduce the impedance presented to the return current by its local return compared to other paths, because the enclosed loop area is smallest for this path (Figure 6.2). This is a major reason for the use of twisted pair cable for data transmission.

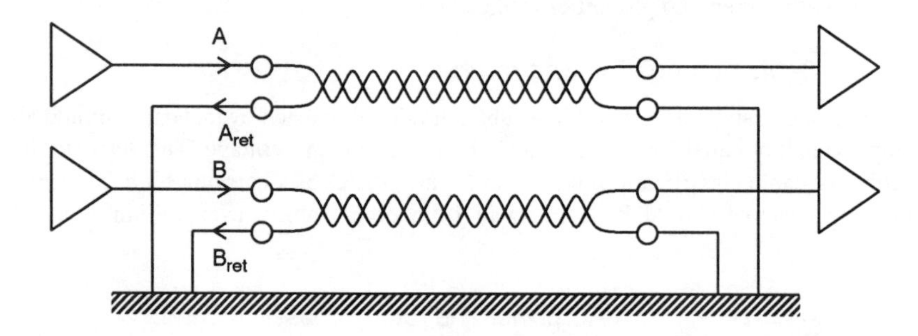

Signal return currents A_{ret} and B_{ret} flow through their local twisted pair return path rather than through ground because this offers the lowest overall path inductance

Figure 6.2 Signal return current paths

This effect is also responsible for the magnetic shielding property of coaxial cable, and is the reason why current in a ground plane remains local to its signal track (compare section 5.1.5).

6.1.2 Cable screens at low frequencies

Optimum screening requires different connection regimes for interference at low frequencies (audio to a few hundred kHz) and at radio frequencies.

6.1.2.1 Screen currents and magnetic shielding

An overall screen, grounded only at one end, provides good shielding from capacitively coupled interference (Figure 6.3(a)) but none at all from magnetic fields, which induce a noise voltage in the loop that is formed when both source and load are grounded. (Beware: different principles apply when either source or load is not grounded!) To shield against a magnetic field, *both* ends of the screen must be grounded. This allows an induced current (I_S in Figure 6.3(b)) to flow in the screen which will negate the current induced in the centre conductor. The effect of this current begins to become apparent only above the cable cut-off frequency, which is a function of the screen inductance and resistance and is around 1 – 2kHz for braided screens or 7 – 10kHz for aluminium foil screens. Above about five times the cut-off frequency, the current induced in the centre conductor is constant with frequency (Figure 6.3(c)).

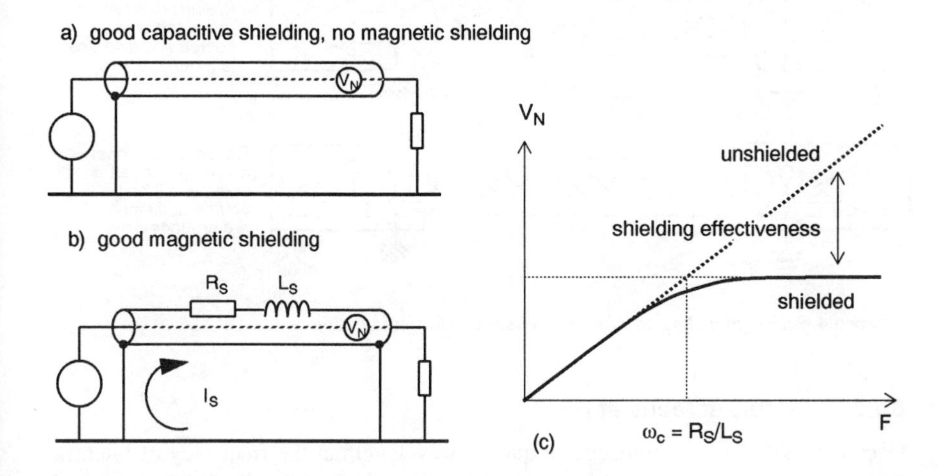

Figure 6.3 Magnetic shielding effectiveness versus screen grounding

The same principle applies when shielding a conductor to prevent magnetic field emission. The return current must flow through the screen, and this will only occur (for a circuit which is grounded at both ends) at frequencies substantially above the shield cut-off frequency.

6.1.2.2 Where to ground the cable screen

The problem with grounding the screen at both ends in the previous circuit is that it becomes a circuit conductor and any voltage dropped across the screen resistance will be injected in series with the signal. Whenever a *circuit* is grounded at both ends, only a limited amount of magnetic shielding is possible because of the large interference currents induced in the screen-ground loop, which develop an interference voltage along the screen. To minimize low frequency magnetic field pickup, one end of the circuit should be isolated from ground, the circuit loop area should be small, and the screen should not form part of the circuit. You can best achieve this by using shielded twisted pair cable with the screen grounded at only one end. The screen then takes care of capacitive coupling while the twisting minimizes magnetic coupling.

For a circuit with an ungrounded source the screen should be grounded at the input common, whereas if the input is floating and the source is grounded then the screen should be grounded to the source common. These arrangements (Figure 6.4) minimize capacitive noise coupling from the screen to the inner conductor(s), since they ensure the minimum voltage differential between the two. Notice though that as the frequency increases, stray capacitance at the nominally ungrounded end reduces the efficiency of either arrangement by allowing undesired ground and screen currents to flow.

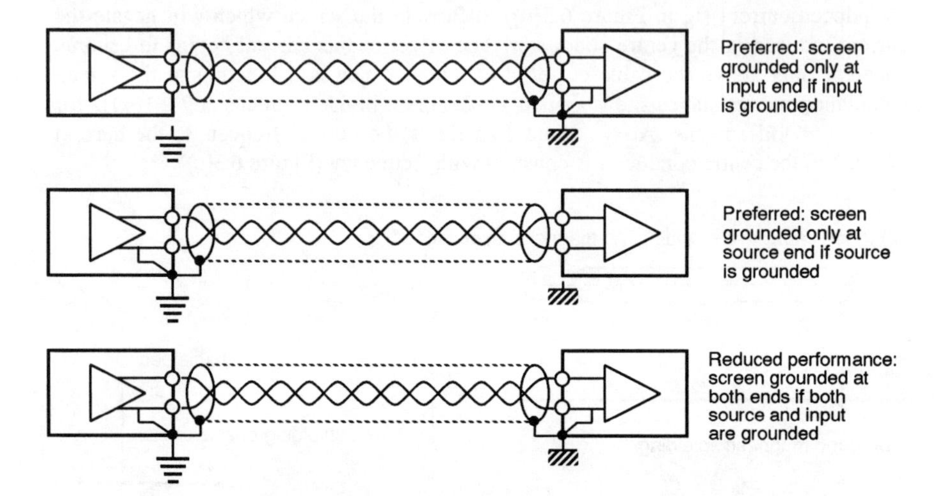

Figure 6.4 Screen grounding arrangements versus circuit configuration

6.1.3 Cable screens at RF

Once the cable length approaches a quarter wavelength at the frequency of interest, screen currents due to external fields become a fact of life. An open circuit at one end of the cable becomes transformed into a short circuit a quarter wavelength away, and screen currents flow in a standing wave pattern whether or not there is an external connection (Figure 6.5). The magnitude of the current is related to the characteristic impedance of the transmission line formed by the cable and the ground plane (this behaviour is discussed in section 4.3.1.1). Even below resonant frequencies, stray capacitance can allow screen currents to flow.

Separation of inner and outer screen currents

However at high frequencies the inner and outer of the screen are isolated by skin effect, which forces currents to remain on the surface of the conductor. Signal currents on the inside of the screen do not couple with interference currents on the outside. Thus multiple grounding of the screen, or grounding at both ends, does not introduce interference voltages on the inside to the same extent as at low frequencies. This effect is compromised by a braided screen due to its incomplete optical coverage and because the strands are continuously woven from inside to out and back again. It is also more seriously compromised by the quality of the screen ground connection at either end, as is discussed in section 6.1.5.

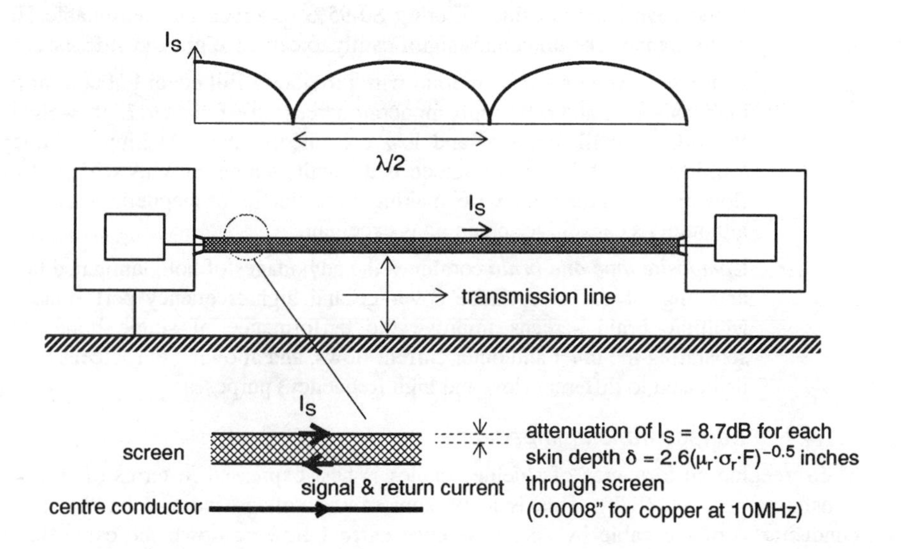

Figure 6.5 The cable screen at RF

6.1.4 Types of cable screen

The performance of cable screens depends on their construction. Figure 6.6 shows some of the more common types of screen available commercially at reasonable cost; for more demanding applications specialized screen constructions such as optimized or multiple braids are available at a premium. Of course, you can also run unscreened cable in shielded conduit, in a separate braided screen or wrap it with screening or permeable material. These options are most useful for systems or installation engineers.

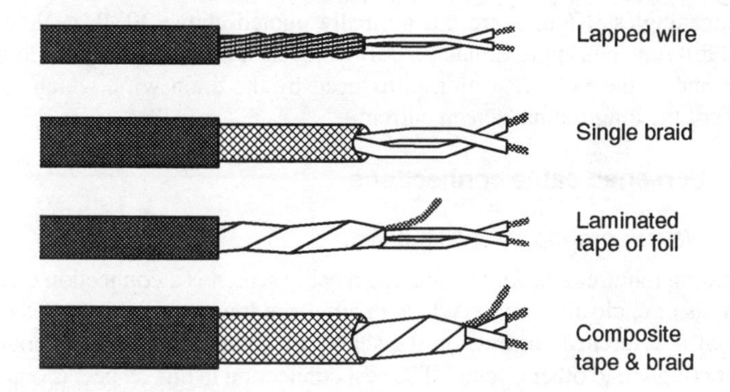

Figure 6.6 Common screen types

- *Lapped wire* screens consist of wires helically wound onto the cable. They are very flexible, but have poor screening effectiveness and are noticeably inductive at HF, so are restricted to audio use.
- *Single braid* screens consist of wire woven into a braid to provide a metallic

frame covering the cable, offering 80-95% coverage and reasonable HF performance. The braid adds significantly to cable weight and stiffness.

- *Laminated tape or foil* with drain wire provides a full cover but at a fairly high resistance and hence only moderate screening efficiency. Light weight, flexibility, small diameter and low cost are retained. Making a proper termination to this type of screen is difficult; screen currents will tend to flow mainly in the drain wire, making it unsuitable for magnetic screening, although its capacitive screening is excellent.
- *Composite tape and braid* combines the advantages of both laminated tape and single braid to optimise coverage and high frequency performance. Multiple braid screens improve the performance of single braids by separating the inner and outer current flows, and allowing the screens to be dedicated to different (low and high frequency) purposes.

6.1.4.1 Surface transfer impedance

The screening performance of shielded cables is best expressed in terms of surface transfer impedance (STI). This is a measure of the voltage induced on the inner conductor(s) of the cable by an interference current flowing down the cable outer shield, which will vary with frequency and is normally expressed in milliohms per metre length. A perfect screen would not allow any voltage to be induced on the inner conductors and would have an STI of zero, but practical screens will couple some energy onto the inner via the screen impedance. At low frequencies it is equal to the dc resistance of the screen.

Figure 6.7 compares STI versus frequency for various types of cable screen construction. The decrease in STI with frequency for the better performance screens is due to the skin effect separating signal currents on the inside of the screen from noise currents on the outside. The subsequent increase is due to field distortion by the holes and weave of the braid. Note that the inexpensive types have a worsening STI with increasing frequency. Once the frequency approaches cable resonance then STI figures become meaningless; figures are not normally quoted above 30MHz. Note that the laminated foil screen is approximately 20dB worse than a single braid, due to its higher resistance and to the field distortion introduced by the drain wire, which carries the major part of the longitudinal screen current.

6.1.5 Screened cable connections

6.1.5.1 How to ground the cable shield

The overriding requirement for terminating a cable screen is a connection direct to the metal chassis or enclosure ground which exhibits the lowest possible impedance. This ensures that interference currents on the shield are routed to ground without passing through or coupling to other circuits. The best connection in this respect is one in which the shield is extended up to and makes a solid 360° connection with the ground plane or chassis (Figure 6.8). This is best achieved with a hard-wired cable termination using a conductive gland and ferrule which clamps over the cable screen. A connector will always compromise the quality of the screen-to-chassis bond, but some connectors are very much better than others.

Connector types

Military-style connectors allow for this construction, as do the standard ranges of RF

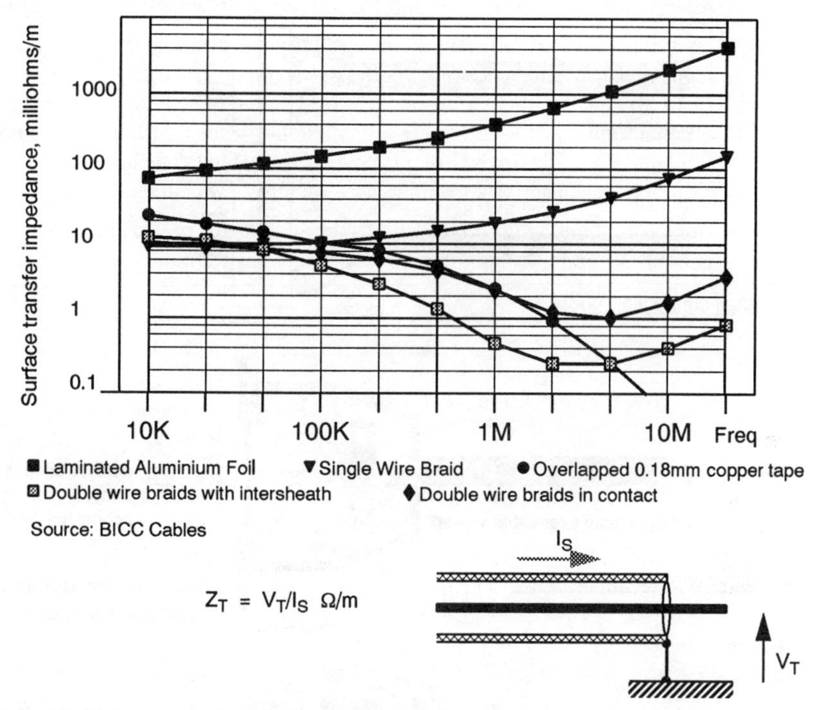

Figure 6.7 Surface transfer impedance of various screen types

coaxial connectors such as N type or BNC. Of the readily available commercial multi-way connectors, only those with a connector shell that is designed to make positive 360° contact with its mate are suitable. Examples are the subminiature D range with dimpled tin-plated shells. Connector manufacturers are now introducing properly designed conductive shells for other ranges of mass-termination connector as well.

The importance of the backshell

The cable screen must make 360° contact with a screened, conductive backshell which must itself be positively connected to the connector shell. The 360° contact is best offered by an iris or ferrule arrangement although a well-made conductive clamp to the backshell body is an acceptable alternative. A floating cable clamp, or a backshell which is not tightly mated to the connector shell are not adequate. The backshell itself can be conductively coated plastic rather than solid metal with little loss of performance, because the effect of the 360° termination is felt at the higher frequencies where the skin depth allows the use of very thin conductive surfaces. On the other hand, the backshell is *not* primarily there to provide electric field screening; simply using a metal or conductively coated shell without ensuring a proper connection to it is pointless.

6.1.5.2 The effect of the pigtail

A pigtail connection is one where the screen is brought down to a single wire and extended through a connector pin to the ground point. Because of its ease of assembly

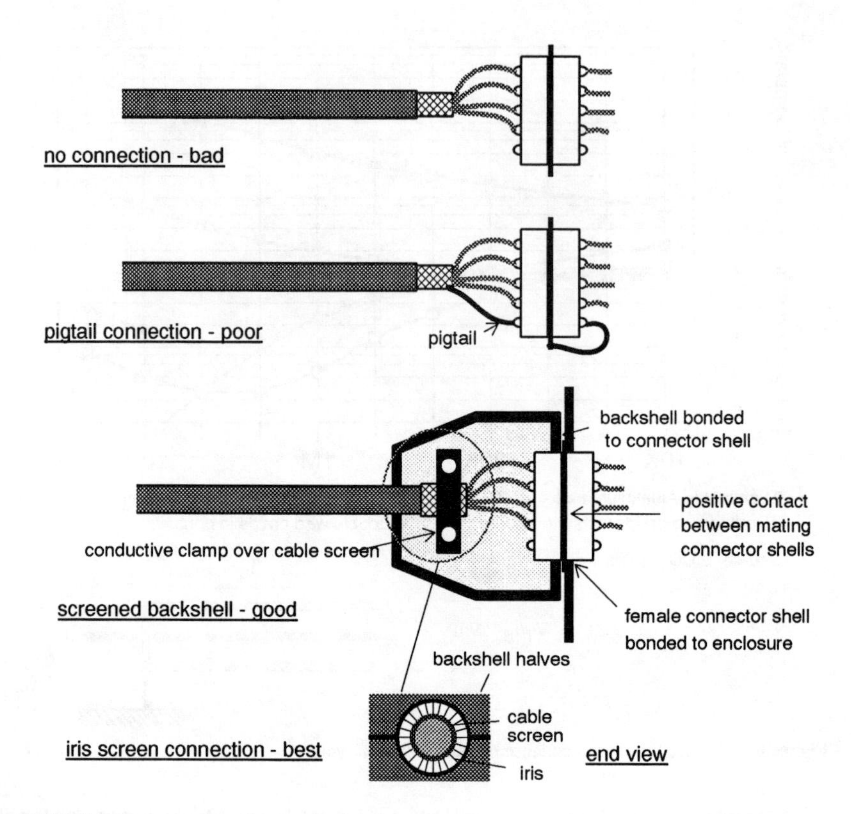

Figure 6.8 Cable screen connection methods

it is very commonly used for connecting the screens of data cables. Unfortunately, it may be almost as bad as no connection at high frequencies because of the pigtail inductance [50][68]. This can be visualised as being a few tens of nanohenries in series with the screen connection (Figure 6.9), which develops a common-mode voltage on the cable screen at the interface as a result of the screen currents.

The equivalent surface transfer impedance of such a connection rises rapidly with increasing frequency until it is dominated by the pigtail inductance, and effectively negates the value of a good HF screened cable. At higher frequencies resonances with the stray capacitances around the interface limit the impedance, but they also make the actual performance of the connection unpredictable and very dependent on construction and movement. If a pigtail connection is unavoidable then it must be as short as possible, and preferably doubled and taken through two pins on opposite ends of the connector so that its inductance is halved.

Effective length

Note that the effective length of the pigtail extends from the end of the cable screen through the connector and up to the point of the ground plane or chassis connection. The common practice of mounting screened connectors on a pcb with the screening shell taken to ground via a length of track – which sometimes travels the length of the board – is equivalent to deliberate insertion of a pigtail on the opposite side of the connection.

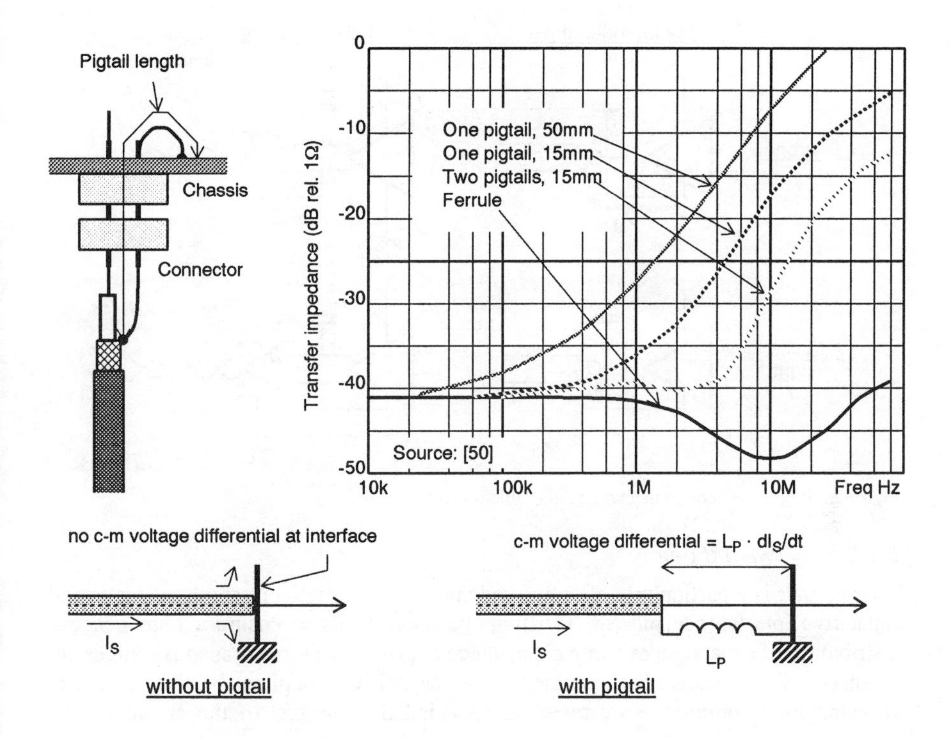

Figure 6.9 The pigtail

Screened connectors must always be mounted so that their shells are bonded directly to chassis.

6.1.5.3 Digital I/O decoupling

Input/output decoupling is of critical importance because it is vital to keep cable common-mode interference currents to a minimum. We have already looked at the optimum configuration for I/O cable connections (section 5.1.7). If the cable screen or return is taken to the wrong point with respect to the output driver decoupling capacitor, the high-speed current transitions on the driver supply (which flow through the decoupling capacitor traces) generate common-mode voltage noise V_N (Figure 6.10) which is delivered to the cable and which appears as a radiated emission. Cable screens must always be taken to a point at which there is the minimum noise with respect to the system's ground reference.

6.1.6 Unscreened cables

You are not always bound to use screened cable to combat EMC problems. The various unscreened types offer major advantages in terms of cost and the welcome freedom from the need to terminate the screen properly. In situations where the cable carries signal circuits that are not in themselves susceptible or emissive, and where common-mode cable currents are inoffensive or can be controlled at the interface by other means such as filtering, unscreened cables are quite satisfactory.

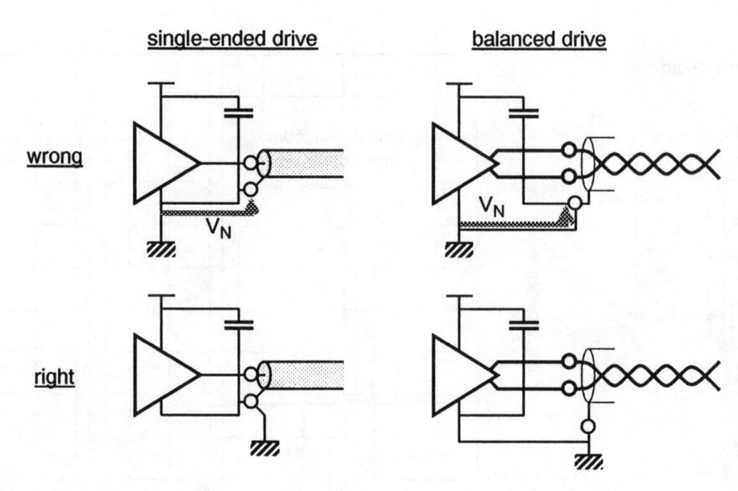

Figure 6.10 The point of connection of I/O cable screens

6.1.6.1 Twisted pair

Twisted pair is a particularly effective and simple way of reducing both magnetic and capacitive interference pickup. Twisting the wires tends to ensure a homogeneous distribution of capacitances. Both capacitance to ground and to extraneous sources are balanced. This means that common-mode capacitive coupling is also balanced, allowing high common-mode rejection provided that the rest of the circuit is also balanced.

Twisting is most useful in reducing low-frequency magnetic pickup because it reduces the magnetic loop area to almost zero. Each twist reverses the direction of induction so, assuming a uniform external field, two successive twists cancel the wires' interaction with the field. Effective loop pickup is now reduced to the small areas at each end of the pair, plus some residual interaction due to non-uniformity of the field and irregularity in the twisting. If the termination area is included in the field, the number of twists per unit length is unimportant [28][31]. Clearly, the un-twisted termination area or length should be minimised. If the field is localised along the cable, performance improves as the number of twists per unit length increases.

6.1.6.2 Ribbon cable

Ribbon is widely used for parallel data transmission within enclosures. It allows mass termination to the connector and is therefore economical. It should be shielded if it carries high frequency signals and is extended outside a screened enclosure, but you will find that proper termination of the shield is usually incompatible with the use of a mass-termination connector. Ribbon cable can be obtained with an integral ground plane underneath the conductors, or with full coverage screening. Figure 6.11 shows the relative merits of each in reducing emissions from a typical digital signal. However, the figures for ground plane and shielded cables assume a low-inductance termination, which is difficult to achieve in practice; typical terminations via drain wires will worsen this performance, more so at high frequencies.

Ground configuration in ribbon

The performance of ribbon cables carrying high frequency data is very susceptible to

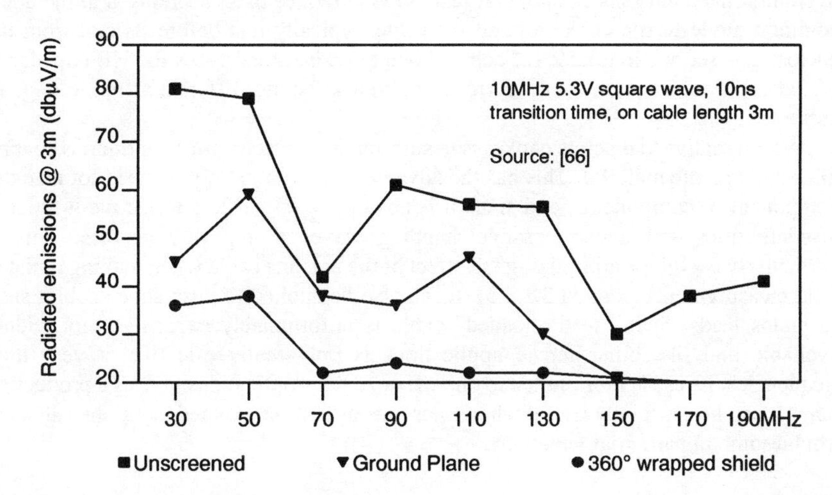

Figure 6.11 Coupling from different types of ribbon cable

the configuration of the ground returns. The cheapest configuration is to use one ground conductor for the whole cable (Figure 6.12(a)). This creates a large inductive loop area for the signals on the opposite side of the cable, and crosstalk and ground impedance coupling between signal circuits. The preferred configuration is a separate ground return for each signal (b). This gives almost as good performance as a properly terminated ground plane cable, and is very much easier to work with. Crosstalk and common impedance coupling is virtually eliminated. Its disadvantage is the extra size and cost of the ribbon and connectors. An acceptable alternative is configuration (c), two signal conductors per return. This improves cable utilization by 50% over (b) and maintains the small inductive loop area, at the expense of possible crosstalk and ground coupling problems. The optimum configuration of (b) can be improved even more by using twisted pair configured into the ribbon construction.

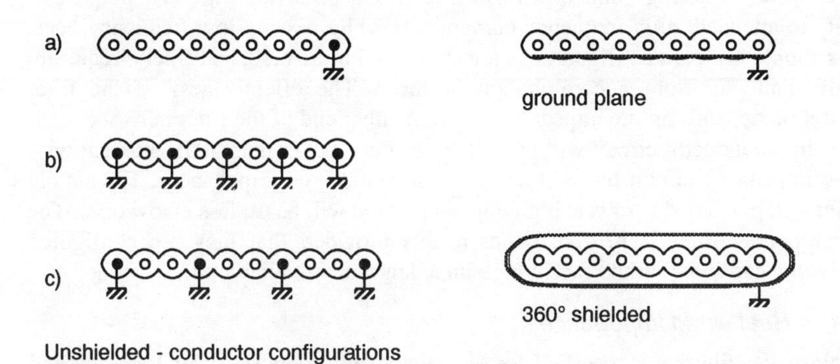

Figure 6.12 Ribbon cable configurations

6.1.6.3 Ferrite loaded cable

Common mode currents in cable screens are responsible for a large proportion of overall radiated emission. A popular technique to reduce these currents is to include a common mode ferrite choke around the cable, typically just before its exit from the enclosure – see section 6.2.2.1. Such a choke effectively increases the HF impedance of the cable to common mode currents without affecting differential mode (signal) currents.

An alternative to discrete chokes is to surround the screen with a continuous coating of flexible ferrite material. This has the advantage of eliminating the need for an extra component or components, and since it is absorptive rather than reflective it reduces discontinuities and hence possible standing waves at high frequencies. This is particularly useful for minimizing the effect of the antenna cable when making radiated field measurements (section 3.1.5.3). It can also be applied to unscreened cables such as mains leads. Such "ferrite-loaded" cable is unfortunately expensive, not widely available and like other ferrite applications is only really effective at very high frequencies. Its use is more suited to one-off or ad hoc applications than as a production item. It can be especially useful when transients or ESD conducted along the cable are troublesome in particular situations.

6.2 Filtering

It is normally impossible to completely eliminate noise being conducted out of or into equipment along connecting leads. The purpose of filtering is to attenuate such noise to a level either at which it meets a given specification, for exported noise, or at which it does not result in malfunction of the system, for imported noise. If a filter contains lossy elements, such as a resistor or ferrite component, then the noise energy may be absorbed and dissipated within the filter. If it does not – i.e. if the elements are purely reactive – then the energy is reflected back to its source and must be dissipated elsewhere in the system. This is one of the features which distinguishes EMI filter design from conventional signal filter design, that in the stop-band the filter should be as lossy as possible.

6.2.1 Filter configuration

In EMC work, "filtering" almost always means low-pass filtering. The purpose is normally to attenuate high frequency components while passing low frequency ones. Various simple low pass configurations are shown in Figure 6.13, and filter circuits are normally made up from a combination of these. The effectiveness of the filter configuration depends on the impedances seen at either end of the filter network.

The simple inductor circuit will give good results – better than 40dB attenuation – in a low impedance circuit but will be quite useless at high impedances. The simple capacitor will give good results at high impedances but will be useless at low ones. The multi-component filters will give better results provided that they are configured correctly; the capacitor should face a high impedance and the inductor a low one.

6.2.1.1 Real world impedances

Conventionally, filters are specified for terminating impedances of 50Ω at each end because this is convenient for measurement and is an accepted RF standard. In the real application, Z_S and Z_L are complex and perhaps unknown at the frequencies of interest for suppression. If either or both has a substantial reactive component then resonances are created which may convert an insertion loss into an insertion gain at some frequencies. Differential mode impedances may be predictable if the components

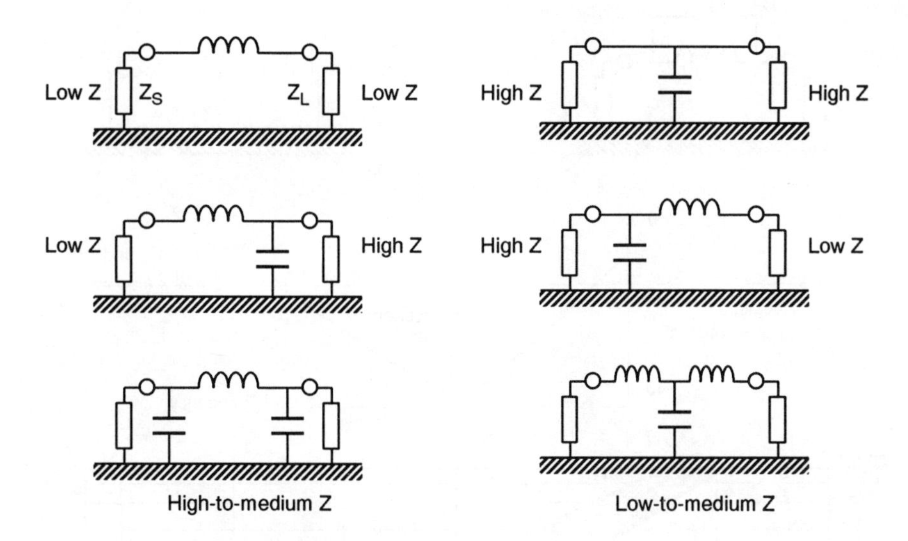

Figure 6.13 Filter configuration versus impedance

which make up the source and load are well characterized at RF, but common mode impedances such as are presented by cables or the stray reactances of mechanical structures are essentially unpredictable. Practically, cables have been found to have common mode impedances in the region of 100 to 400Ω except at resonance, and a figure of 150Ω is commonly taken for a rule of thumb (see also section 4.3.1.1).

6.2.1.2 Parasitic reactances

Filter components, like all others, are imperfect. Inductors have self-capacitance, capacitors have self-inductance. This complicates the equivalent circuit at high frequencies and means that a typical filter using discrete components will start to lose its performance above about 10MHz. The larger the components are physically, the lower will be the break frequency. For capacitors, as the frequency increases beyond capacitor self-resonance the impedance of the capacitors in the circuit actually rises, so that the insertion loss begins to fall. This can be countered by using special construction for the capacitors (see the next section). Similarly, inductors have a self-resonant frequency beyond which their impedance starts to fall. Filter circuits using a single choke are normally limited in their performance by the self resonance of the choke (see Figure 6.14) to 40 or 50dB. Better performance than this requires multiple filter sections

Capacitors

Ceramic capacitors are usually regarded as the best for RF purposes, but in fact the subminiature polyester or polystyrene film types are often perfectly adequate, since their size approaches that of the ceramic type. Small capacitors with short leads (the ideal is a chip component) will have the lowest self inductance. For EMI filtering, lossy dielectrics such as X7R, Y5V and Z5U are an advantage.

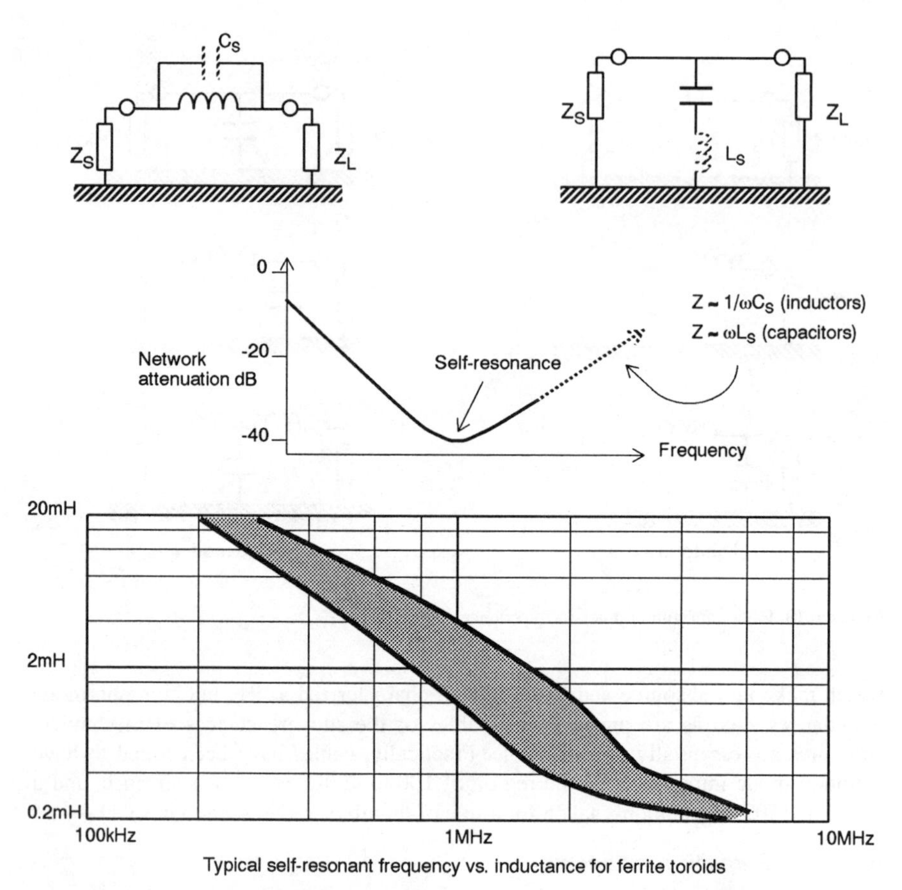

Figure 6.14 Self resonant effects due to parasitic reactances [32]

Inductors

The more turns an inductor has, the higher will be its inductance but also the higher its self capacitance. The number of turns for a given inductance can be reduced by using a high permeability core, but these also exhibit a high dielectric constant which tends to increase the capacitance again, for which reason you should always use a bobbin on a high permeability core rather than winding directly onto the core. For minimum self capacitance the start and finish of a winding should be widely separated; winding in sections on a multi-section bobbin is one way to achieve this. A single layer winding exhibits the lowest self capacitance. If you have to use more turns than can be accommodated in a single layer, progressive rather than layer winding (see Figure 6.15) will minimize the capacitance.

6.2.1.3 Component layout

Lead inductance and stray capacitance degrade filter performance markedly at high frequency. Two common faults in filter applications are not to provide a low-inductance ground connection, and to wire the input and output leads in the same loom or at least close to or passing each other. This construction will offer low frequency

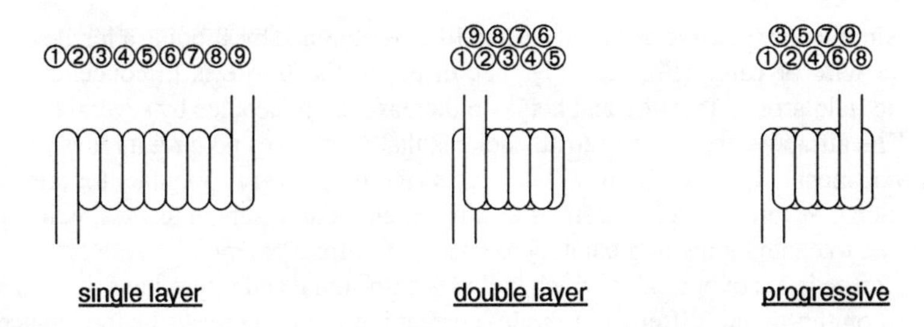

Figure 6.15 Inductor winding techniques

differential mode attenuation but high frequency common mode attenuation will be minimal.

A poor ground offers a common impedance which rises with frequency and couples HF interference straight through via the filter's local ground path. Common input-output wiring does the same thing through stray capacitance or mutual inductance, and it is also possible for the "clean" wiring to couple with the unfiltered side through inappropriate routing. The cures (Figure 6.16) are to mount the filter so that its ground node is directly coupled to the lowest inductance ground of the equipment, preferably the chassis, and to keep the I/O leads separate, preferably screened from each other. The best solution is to position the filter so that it straddles the equipment shielding, where this exists.

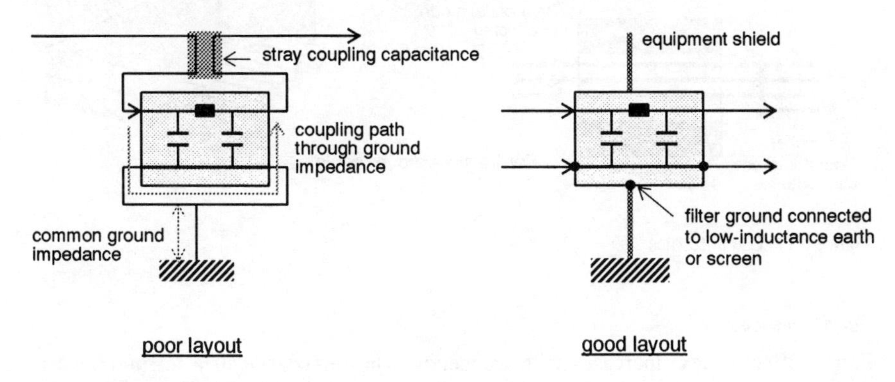

Figure 6.16 The effect of filter layout

Component layout within the filter itself is also important. Input and output components should be well separated from each other for minimum coupling capacitance, while all tracks and in particular the ground track should be short and substantial. It is best to lay out the filter components exactly as they are drawn on the circuit diagram.

6.2.2 Components

There are a number of specialized components which are intended for EMI filtering applications.

6.2.2.1 Ferrites

A very simple, inexpensive and easily-fitted filter is obtained by slipping a ferrite sleeve around a wire or cable (Figure 6.17). The effect of the ferrite is to concentrate the magnetic field around the wire and hence to increase its inductance by several hundred times. The attractiveness of the ferrite choke is that it involves no circuit redesign, and often no mechanical redesign either. It is therefore very popular for retro-fit applications. Several manufacturers offer kits which include halved ferrites, which can be applied to cable looms immediately to check for improvement.

If a ferrite is put over a cable which includes both signal and return lines, it will have no effect on the signal (differential-mode) current but it will increase the impedance to common-mode currents. The effectiveness can be increased by looping the cable several times through the core, or by using several cores in series. Stray capacitance limits the improvement that can be obtained by extra turns.

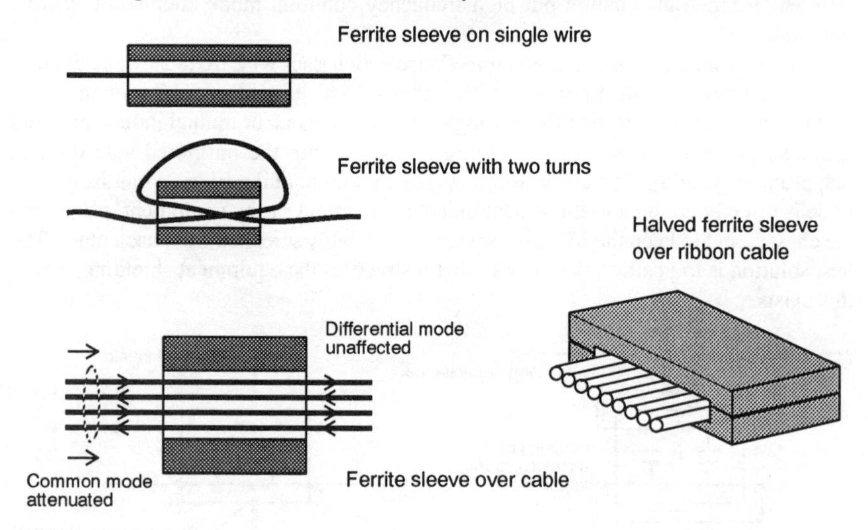

Figure 6.17 Use of ferrites

Ferrite impedance

Ferrite effectiveness increases with frequency. The impedance of a ferrite choke is typically around 50Ω at 10MHz, rising to hundreds of ohms above 100MHz (the actual value depends on shape and size, more ferrite giving more impedance). The impedance variation with frequency differs to some extent between manufacturers and between different grades of ferrite material. Figure 6.18 shows this for two grades of ferrite with the same geometry (a bead of 0.195" OD and 0.432" length).

A useful property of ferrites is that they are lossy at high frequencies, so that interference energy tends to be absorbed rather than reflected. This reduces the Q of inductive suppressor circuits and minimizes resonance problems. In fact, ferrites for suppression purposes are especially developed to have high losses, in contrast to those for low frequency or power inductors which are required to have minimum losses. Figure 6.19 shows how the resistive component dominates the impedance characteristic at the higher frequencies.

Because a ferrite choke is no more than a lossy inductor, it only functions usefully

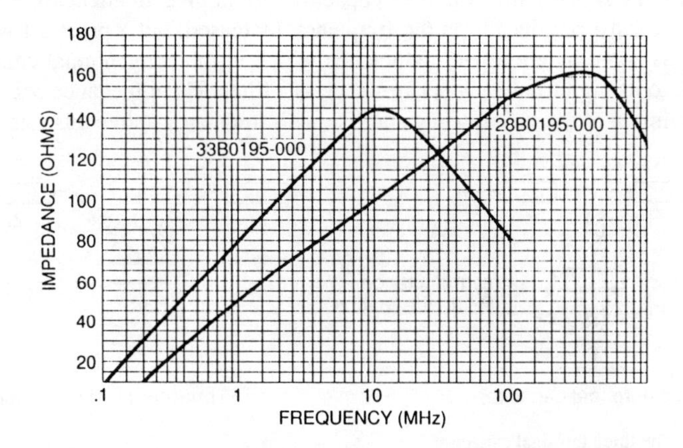

Figure 6.18 Impedance versus frequency for two grades of material (Source: Steward)

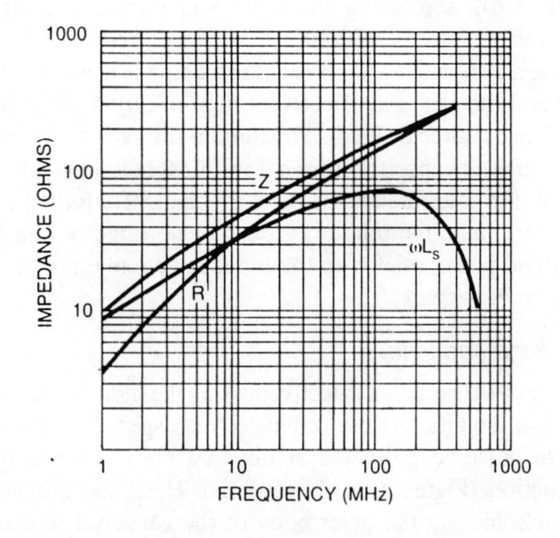

Figure 6.19 Impedance components versus frequency (Source: Steward)

between low impedances. A ferrite included in a high-impedance line will offer little or no attenuation (the attenuation can be derived from the equivalent circuit of Figure 6.13 on page 183). Most circuits, and especially cables, show impedances that vary with frequency in a complex fashion but normally stay within the bounds of 10 – 1000Ω, so a single ferrite will give modest attenuation factors averaging around 10dB and rarely better than 20dB.

A ferrite choke is particularly effective at slowing the fast rate-of-rise of an electrostatic discharge current pulse which may be induced on internal cables. The transient energy is absorbed in the ferrite material rather than being diverted or reflected to another part of the system.

6.2.2.2 Three-terminal capacitors

Any low-pass filter configuration except for the simple inductor uses a capacitor in parallel with the signal path. A perfect capacitor would give an attenuation increasing at a constant 20dB per decade as the frequency increased, but a practical wire-ended capacitor has some inherent lead inductance which in the conventional configuration puts a limit to its high frequency performance as a filter. The impedance characteristics show a minimum at some frequency and rise with frequency above this minimum.

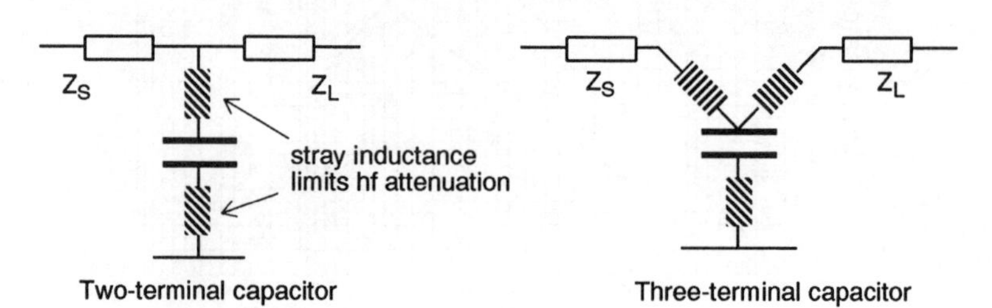

Figure 6.20 The three terminal capacitor

This lead inductance can be put to some use if the capacitor is given a three-terminal construction (Figure 6.20), separating the input and output connections. The lead inductance now forms a T-filter with the capacitor, greatly improving its high-frequency performance. A ferrite bead on each of the upper leads will further enhance the lead inductance and increase the effectiveness of the filter when it is used with a relatively low impedance source or load.. The three-terminal configuration can extend the range of a small ceramic capacitor from below 50MHz to beyond 200MHz, which is particularly useful for interference in the vhf band. To fully benefit from this approach, you must terminate the middle (ground) lead directly to a low inductance ground such as a ground plane, otherwise the inductance remaining in this connection will defeat the capacitor's purpose.

6.2.2.3 Feedthrough capacitors

Any leaded capacitor is still limited in effectiveness by the inductance of the connection to the ground point. For the ultimate performance, and especially where penetration of a screened enclosure must be protected at uhf and above then a feedthrough (or leadthrough) construction (Figure 6.21) is essential. Here, the ground connection is made by screwing or soldering the outer body of the capacitor directly to the metal screening or bulkhead. Because the current to ground can spread out for 360° around the central conductor, there is effectively no inductance associated with this terminal and the capacitor performance is maintained well into the GHz region. This performance is compromised if a 360° connection is not made or if the bulkhead is limited in extent. The inductance of the through lead can be increased, thereby creating a π-section filter, by separating the ceramic metallization into two parts and incorporating a ferrite bead within the construction. Feedthrough capacitors are available in a wide range of voltage and capacitance ratings but their cost increases with size.

6.2.2.4 Chip capacitors

Although they are not seen principally as EMI filter components, surface mounting chip capacitors offer an extra advantage for this use, which is that their lead inductance is zero. The overall inductance is reduced to that of the component itself, which is typically three to five times less than the lead plus component inductance of a

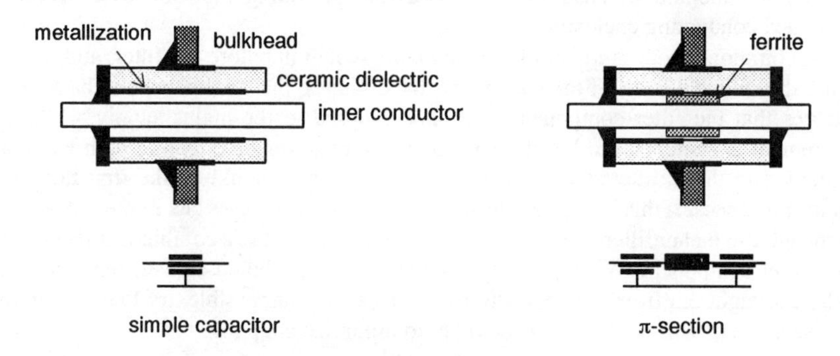

Figure 6.21 The feedthrough capacitor

conventional part. Their self resonant frequency can therefore be double that of a leaded capacitor of the same value. Tracks to capacitors used for filtering and decoupling should be short and direct, in order not to lose this advantage through additional track inductance.

6.2.3 Mains filters

RFI filters for mains supply inputs have developed as a separate species and are available in many physical and electrical forms from several specialist manufacturers. A typical "block" filter for European mains supplies with average insertion loss might cost around £5. Some of the reasons for the development and use of block mains filters are:

- Mandatory conducted emission standards concentrate on the mains port, hence there is an established market for filter units
- Add-on "fit and forget" filters can be retro-fitted
- Safety approvals for the filter have already been achieved
- Many equipment designers are not familiar with RF filter design

In fact, the market for mains filters really took off with the introduction of VDE and FCC standards regulating conducted mains emissions, compounded by the rising popularity of the switchmode power supply. With a switching supply, a mains filter is essential to meet these regulations. EMC has historically tended to be seen as an afterthought on commercial equipment, and there have been many occasions on which retro-fitting a single component mains filter has brought a product into compliance, and this has also encouraged the development of the mains filter market. A real benefit is that safety approvals needed for all components on the mains side of the equipment have been already dealt with by the filter manufacturer if a single-unit filter is used.

6.2.3.1 Application of mains filters

Merely adding a block filter to a mains input will improve low frequency emissions such as the low harmonics of a switching power supply. But HF emissions (above 1MHz) require attention to the layout of the circuitry around the filter (see section 6.2.1.3). Treating it like any other power supply component will not give good HF attenuation and may actually worsen the coupling, through the addition of spurious resonances and coupling paths. Combined filter and CEE22 inlet connector modules are a good method of ensuring correct layout, providing they are used within a grounded conducting enclosure.

A common layout fault is to wire the mains switch in before the filter, and then to bring the switch wiring all the way across the circuit to the front panel and back. This ensures that the filter components are only exposed to the mains supply while the equipment is switched on, but it also provides a ready-made coupling path via stray induction to the unfiltered wiring. Preferably, the filter should be the first thing the mains input sees. If this is impossible, then mount switches, fuses etc. immediately next to the inlet so that unfiltered wiring lengths are minimal, or use a combined inlet/switch/fuse/filter component. Wiring on either side of the filter should be well separated and extend straight out from the connections. If this also is impossible, try to maintain the two sections of wiring at 90° to each other to minimize coupling.

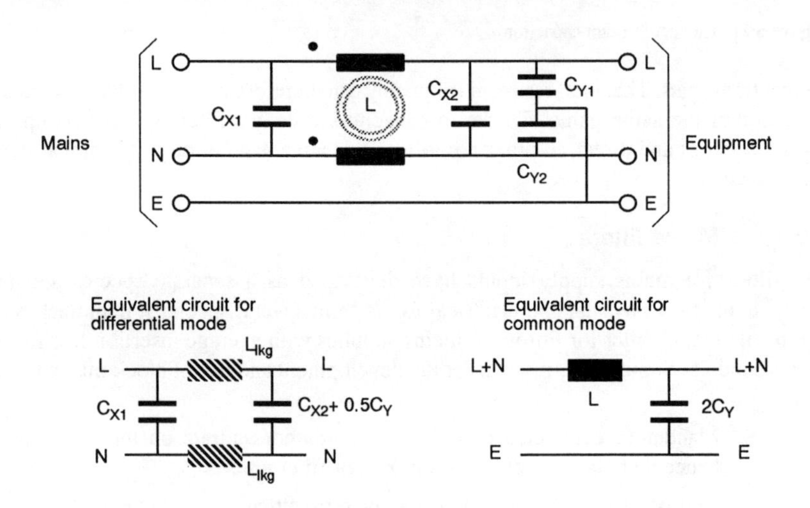

Figure 6.22 Typical mains filter and its equivalent circuit

6.2.3.2 *Typical mains filter*

A typical filter (Figure 6.22) includes components to block both common mode and differential mode components. The common mode choke L consists of two identical windings on a single high permeability, usually toroidal, core, configured so that differential (line-to-neutral) currents cancel each other. This allows high inductance values, typically 1–10mH, in a small volume without fear of choke saturation caused by the mains frequency supply current. The full inductance of each winding is available to attenuate common mode currents with respect to earth, but only the leakage inductance L_{lkg} will attenuate differential mode interference. The performance of the filter in differential mode is therefore severely affected by the method of construction of the choke, since this determines the leakage inductance. A high L_{lkg} will offer greater attenuation, but at the expense of a lower saturation current of the core. Low L_{lkg} is achieved by bifilar winding but safety requirements normally preclude this, dictating a minimum separation gap between the windings.

Common mode capacitors

Capacitors C_{Y1} and C_{Y2} attenuate common mode interference and if C_{X2} is large, have no significant effect on differential mode. The effectiveness of the C_Y capacitors depends very much on the common mode source impedance of the equipment (Figure 6.23). This is usually a function of stray capacitance coupling to earth which depends critically on the mechanical layout of the circuit and the primary-to-secondary capacitance of the mains transformer, and can easily exceed 1000pF. The attenuation offered by the potential divider effect of C_Y may be no more than 15–20dB. The common mode choke is the more effective component, and in cases where C_Y is very severely limited more than one common mode choke may be needed. Calculation of appropriate component values is covered in Appendix C (section C.6).

Differential mode capacitors

Capacitors C_{X1} and C_{X2} attenuate differential mode only but can have fairly high values, 0.1 to 0.47μF being typical. Either may be omitted depending on the detailed performance required, remembering that the source and load impedances may be too low for the capacitor to be useful. For example a 0.1μF capacitor has an impedance of about 10Ω at 150kHz, and the differential mode source impedance seen by C_{X2} may be considerably less than this for a power supply in the hundreds of watts range, so that a C_{X2} of this value would have no effect at the lower end of the frequency range where it is most needed.

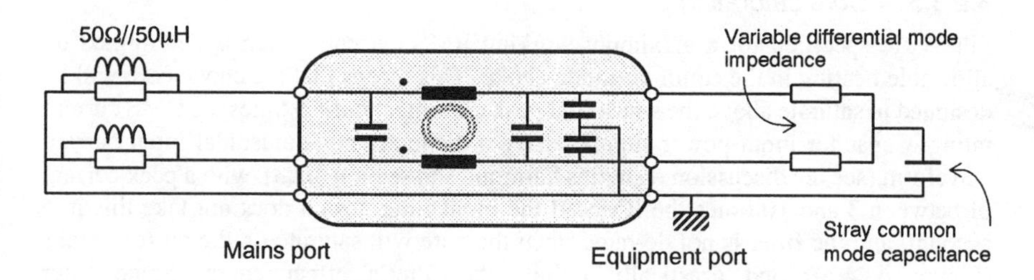

Figure 6.23 Impedances seen by the mains filter

6.2.3.3 Safety considerations

C_{Y1} and C_{Y2} are limited in value by the permissible continuous current which may flow in the safety earth, due to the mains operating voltage impressed across C_{Y1} (or C_{Y2} under certain fault conditions). Values for this current range from 0.25mA to 5mA depending on the approvals authority, safety class and use of the apparatus. Medical equipment has an even lower leakage requirement, typically 0.1mA. Note that this is the *total* leakage current due to the apparatus; if there are other components (such as transient suppressors) which also form a leakage path to earth, the current due to them must be added to that due to C_Y, putting a further constraint on the value of C_Y.

BS 613, which specifies EMI filters in the UK, allows a maximum value for C_Y of 5000pF with a tolerance of ±20% for safety class I or II apparatus, so this value is frequently found in general-purpose filter units.

Both C_X and C_Y carry mains voltages continuously and must be specifically rated to do this. Failure of C_X will result in a fire hazard, while failure of C_Y will result in both a fire hazard and a potential shock hazard. "X" and "Y" class components to BS 6201 part 3 (IEC384-14) are marketed specifically for these positions.

6.2.3.4 *Insertion loss versus impedance*

Ready-made filters are universally specified between 50Ω source and load impedances. The typical filter configuration outlined above is capable of 40-50dB attenuation up to 30MHz in both common and differential modes. Above 30MHz stray component reactances limit the achievable loss and also make it more difficult to predict behaviour. Below 1MHz the attenuation falls off substantially as the effectiveness of the components reduces.

The 50Ω termination does not reflect the real situation. The mains port HF impedance can be generalized for both common and differential mode by a 50Ω//50μH network as provided by a CISPR 16 LISN (section 3.1.2.2); when the product is tested for compliance, this network will be used anyway. The equipment port impedance will vary substantially depending on load and on the HF characteristics of the input components such as the mains transformer, diodes and reservoir. Differential mode impedance is typically a few ohms for small electronic products, while common mode impedance as discussed above can normally be approximated by a capacitive reactance of around 1000pF. The effect of these load impedances differing from the nominal may be to enhance resonances within the filter and thus to achieve insertion gain at some frequencies.

6.2.3.5 *Core saturation*

Filters are specified for a maximum working RMS current, which is determined by allowable heating in the common-mode choke. This means that the choke core will be designed to saturate above the sinusoidal peak current, about 1.5 times the RMS current rating. Capacitor input power supplies have a distinctly non-sinusoidal input current waveform (see the discussion on mains harmonics in section 4.2.3), with a peak current of between 3 and 10 times the RMS. If the input filter design does not take this into account, and the filter is not de-rated, then the core will saturate on the current peaks (Figure 6.24(a)) and drastically reduce the filter's effectiveness. Some filter manufacturers now take this into account and over-rate the inductor, but the allowable peak current is rarely specified on data sheets.

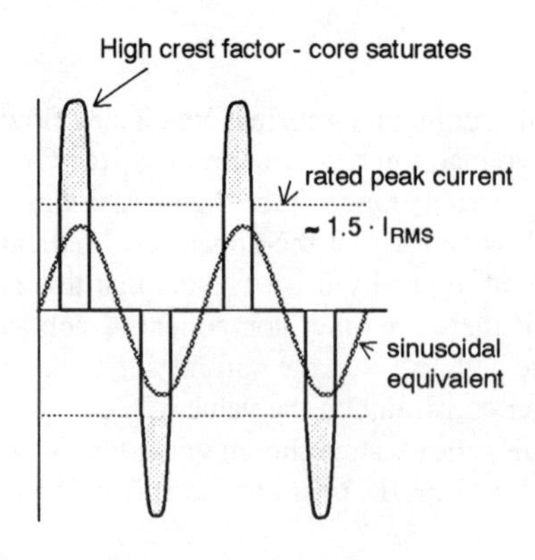

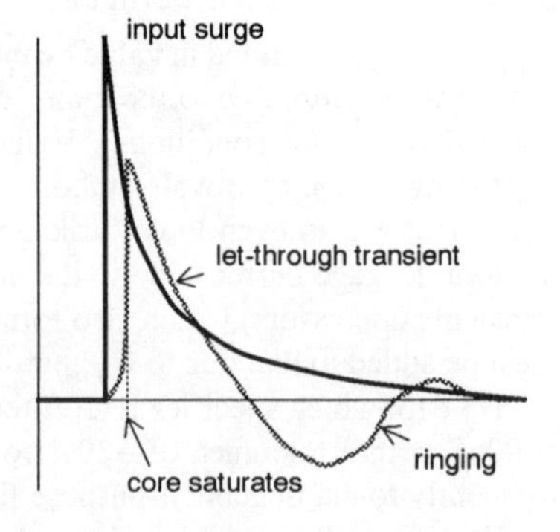

Figure 6.24 Core saturation effects

The core will also saturate when it is presented with a high voltage, high energy common mode surge, such as a switching transient on the mains (Figure 6.24(b)). The surge voltage will be let through delayed and with a slower risetime but only slightly attenuated with attendant ringing on the trailing edge. Standard mains filters designed only for attenuating frequency-domain emissions are inadequate to cope with large incoming common mode transients, though some are better than others. Differential mode transients require considerably more energy to saturate the core and these are more satisfactorily suppressed.

6.2.3.6 Extended performance

In some cases the insertion loss offered by the typical configuration won't be adequate. This may be the case when for example a high power switching supply must meet the most stringent emission limits, or there is excessive coupling of common mode interference, or greater incoming transient immunity is needed. The basic filter design can be extended in a number of ways (Figure 6.25):

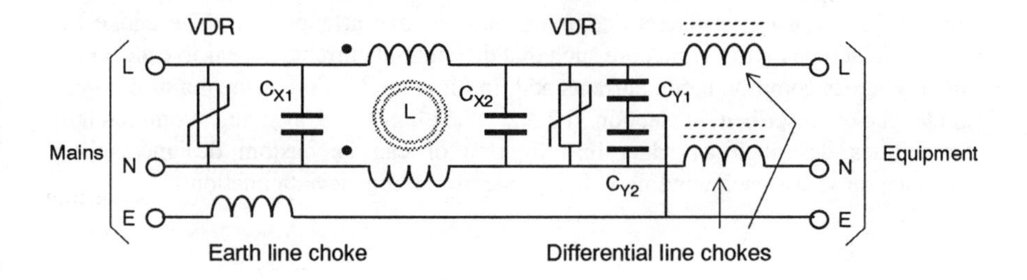

Figure 6.25 Higher performance mains filter

- *extra differential line chokes*: these are separate chokes in L and N lines which are not cross-coupled and therefore present a higher impedance to differential mode signals, giving better attenuation in conjunction with C_X. Because they must not saturate at the full ac line current they are much larger and heavier for a given inductance.
- *an earth line choke*: This increases the impedance to common mode currents flowing in the safety earth and may be the only way of dealing with common mode interference, both incoming and outgoing, when C_Y is already at its maximum limit and nothing can be done about the interference at source. Because it is in series with the safety earth its fault current carrying capability must satisfy safety standards. Ensure that it is not short circuited by an extra earth connection to the equipment case.
- *transient suppressors:* A device such as a voltage dependent resistor (VDR) across L and N will clip incoming differential mode surges (see also section 6.2.5). If it is placed at the mains port then it must be rated for the full expected transient energy, but it will prevent the choke from saturating and protect the filter's C_X; if it is placed on the equipment side then it can be substantially downrated since it is protected by the impedance of the filter. Note that a VDR in these positions has no effect on common mode

transients.

In addition to these extra techniques the basic filter π-section can be cascaded with further similar sections, perhaps with inter-section screens and feedthroughs to obtain much higher insertion loss. For these levels of performance the filter must be used in conjunction with a well screened enclosure to prevent high frequency coupling around it. Large values of C_X should be protected with a bleeder resistor in parallel, to prevent a hazardous charge remaining between L and N when the power is removed (detailed requirements can be found in safety specifications such as IEC 335/EN 60 335).

6.2.4 I/O filtering

If I/O connections carry only low bandwidth signals and low current it is possible to filter them using simple RC low pass networks (Figure 6.26). The decoupling capacitor must be connected to the clean I/O ground (see section 5.1.7) which may not be the same as circuit 0V.

This is not possible with high speed data links, but it is possible to attenuate common mode currents entering or leaving the equipment without affecting the signal frequencies by using a discrete common mode choke arrangement. The choke has several windings on the same core such that differential currents appear to cancel each other whereas common mode currents add, in the same fashion as the mains common mode choke described in section 6.2.3.2. Such units are available commercially (sometimes described as "data line filters") or can be custom designed. Stray capacitance across each winding will degrade high frequency attenuation.

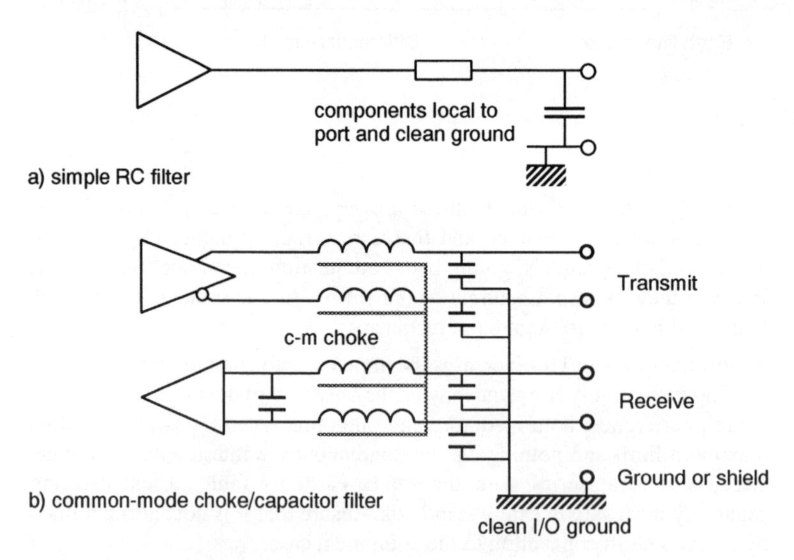

Figure 6.26 I/O filtering techniques

6.2.4.1 Filtered connectors

A convenient way to incorporate both the capacitors and to a lesser extent the inductive components of Figure 6.26 is within the external connector itself. Each pin can be configured as a feedthrough capacitor with a ceramic jacket, its outside metallization connected to a matrix which is grounded directly to the connector shell (Figure 6.27).

Thus the inductance of the ground connection is minimized, provided that the connector shell itself is correctly bonded to a clean ground, normally the metal backplate of the unit. Any series impedance in the ground path not only degrades the filtering but will also couple signals from one line into another, leading to designed-in crosstalk.

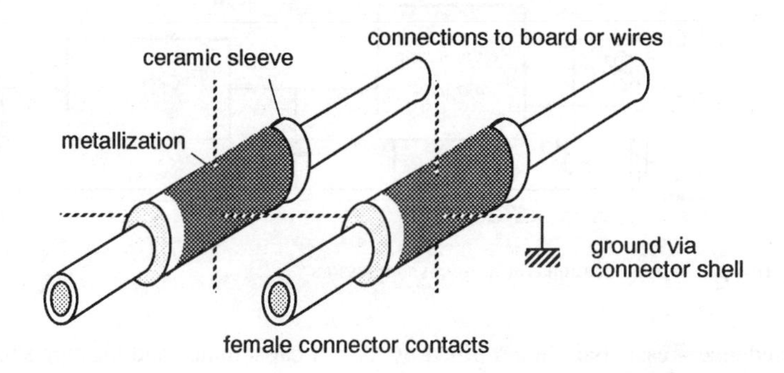

Figure 6.27 Filtered connector pins

The advantage of this construction is that the insertion loss can extend to over 1GHz, the low frequency loss depending entirely on the actual capacitance (typically 50 - 2000pF) inserted in parallel with each contact. With some ferrite incorporated as part of the construction, a π-filter can be formed as with the conventional feedthrough (section 6.2.2.3). No extra space for filtering needs to be provided. The filtered connector has obvious attractions for retro-fit purposes, and may frequently solve interface problems at a stroke. You can also now obtain ferrite blocks tailored to the pinout dimensions of common multiway connectors, which effectively offer individual choking for each line with a single component.

Disadvantages are the significant extra cost over an unfiltered connector; if not all contacts are filtered, or different contacts need different capacitor values, you will need a custom part. Voltage ratings may be barely adequate and reliability may be worsened. Its insertion loss performance at low to medium frequencies can be approached with a small "piggy-back" board of chip capacitors mounted immediately next to the connector.

6.2.4.2 Circuit effects of filtering

When using any form of capacitive filtering, the circuit must be able to handle the extra capacitance to ground, particularly when filtering an isolated circuit at radio frequencies. Apart from reducing the available circuit signal bandwidth, the RF filter capacitance provides a ready-made ac path to ground for the signal circuit and will seriously degrade the ac isolation, to such an extent that an RF filter may actually increase susceptibility to lower frequency common mode interference. This is a result of the capacitance imbalance between the isolated signal and return lines, and it may restrict the allowable RF filter capacitance to a few tens of picofarads.

Capacitive loading of low frequency analogue amplifier outputs may also push the output stage into instability (see section 5.2.3.3).

6.2.5 Transient suppression

Incoming transients on either mains or signal lines are reduced by non-linear devices:

the most common are varistors (voltage dependent resistors, or VDRs), zeners and spark gaps (gas discharge tubes). The device is placed in parallel with the line to be protected (Figure 6.28) and to normal signal or power levels it appears as a high

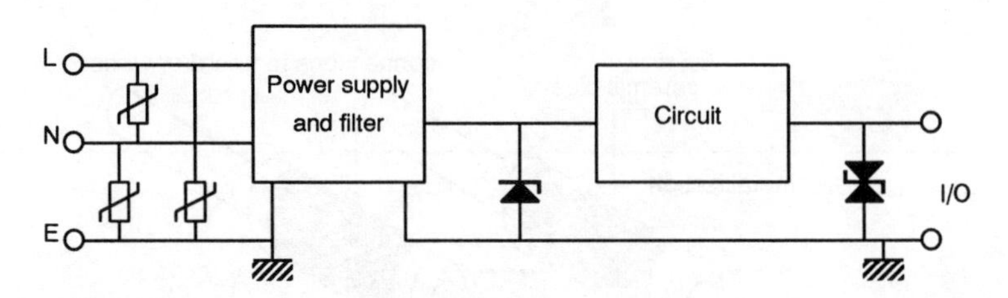

Figure 6.28 Typical locations for transient suppressors

impedance – essentially determined by its self capacitance and leakage specifications. When a transient which exceeds its breakdown voltage appears, the device changes to a low impedance which diverts current from the transient source away from the protected circuit, limiting the transient voltage (Figure 6.29). It must be sized to withstand the continuous operating voltage of the circuit, with a safety margin, and to be able to absorb the energy from any expected transient.

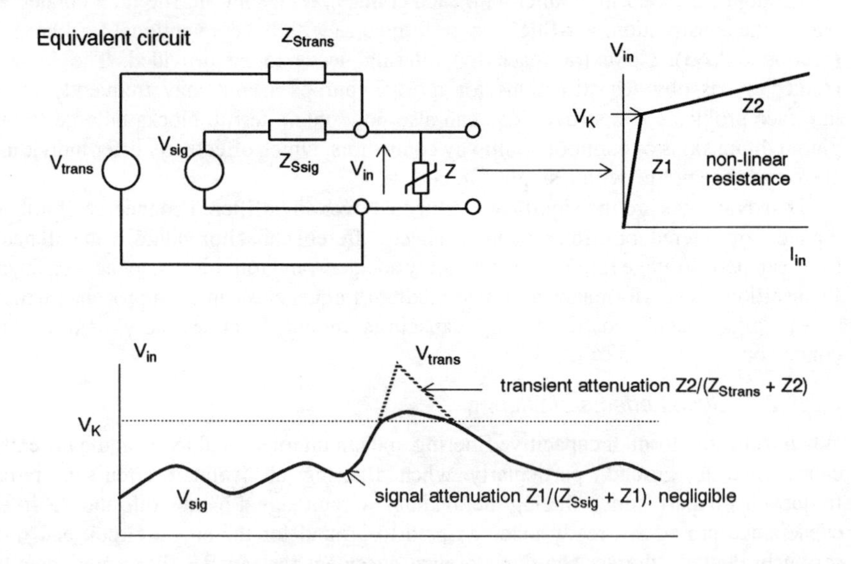

Figure 6.29 The operation of a transient suppressor

The first requirement is fairly simple to design to, although it means that the transient clamping voltage is usually 1.5 – 2 times the continuous voltage, and circuits that are protected by the suppressor must be able to withstand this. The second requirement calls for a knowledge of the source impedance Z_{Strans} and probable amplitude of the transients, which is often difficult to predict accurately especially for external connections. This determines the amount of energy which the suppressor will have to absorb. [29] gives details of how to determine the required suppressor characteristics from a knowledge of the circuit parameters, and also suggests design values for the energy requirement for suppressors on ac power supplies. These are summarized in Table 6.1.

Type of Location	Waveform	Amplitude	Energy deposited in a suppressor with clamping voltage of 500V (120V system)	1000V (240V system)
Long branch circuits and outlets	0.5μs/100kHz oscillatory	6kV/200A	0.8J	1.6J
Major feeders and short branch circuits	0.5μs/100kHz oscillatory	6kV/500A	2J	4J
	8/20ms surge	6kV/3kA	40J	80J

Table 6.1 Suggested transient suppressor design parameters

Table 6.2 compares the characteristics of the most common varieties of transient suppressor.

Device	Leakage	Follow-on current	Clamp voltage	Energy capability	Capacitance	Response time	Cost
ZnO varistor	Moderate	No	Medium	High	High	Medium	Low
Zener	Low	No	Low to medium	Low	Low	Fast	Moderate
Spark gap GDT	Zero	Yes	High ignition, low clamp	High	Very low	Slow	Moderate to high

Table 6.2 Comparison of transient suppressor types

Combining types

You may sometimes have to parallel different types of suppressor in order to achieve a performance which could not be given by one type alone. The disadvantages of straightforward zener suppressors, that their energy handling capability is limited because they must dissipate the full transient current at their breakdown voltage, are overcome by a family of related suppressors which integrate a thyristor with a zener. When the overvoltage breaks down the zener, the thyristor conducts and limits the applied voltage to a low value, so that the power dissipated is low and a given package can handle about ten times the current of a zener on its own. Provided that the operating circuit current is less than the thyristor holding current, the thyristor stops conducting once the transient has passed.

6.2.5.1 Layout of transient suppressors

Short and direct connections to the suppressor (including the ground return path) are vital to avoid compromising the high-speed performance by undesired extra inductance. Transient edges have very fast risetimes (a few nanoseconds for switching-induced interference down to sub-nanosecond for ESD) and any inductance in the clamping circuit will generate a high differential voltage during the transient plus ringing after it, which will defeat the purpose of the suppressor.

The component leads must be short (suppressors are available in SM chip form) and they must be connected locally to the circuit that is to be clamped (Figure 6.30). Any common impedance coupling, via ground or otherwise, must be avoided. When the expected transient source impedance is low (less than a few ohms), it is worthwhile raising the RF impedance of the input circuit with a lossy component such as a ferrite bead. Where suppressors are to be combined with I/O filtering you may be able to use the 3-terminal varistor/capacitor devices that are now available.

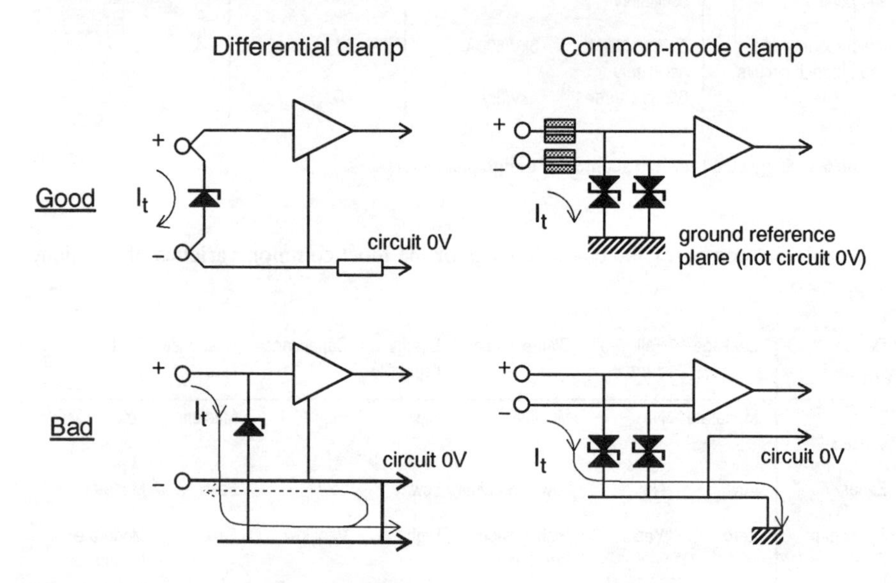

Figure 6.30 Layout and configuration of I/O transient suppressors

6.2.6 Contact suppression

An opening contact which interrupts a flow of current – typically a switch or relay – will initiate an arc across the contact gap. The arc will continue until the available current is not enough to sustain a voltage across the gap (Figure 6.31). The stray capacitance and inductance associated with the contacts and their circuit will in practice cause a repetitive discharge until their energy is exhausted, and this is responsible for considerable broadband interference [10][46]. A closure can also cause interference because of contact bounce.

Any spark-capable contact should be suppressed. The criteria for spark capability are a voltage across the contacts of greater than 320V, and/or a circuit impedance which allows a dV/dt of greater than typically 1V/μs – this latter criterion being met by many low-voltage circuits. The conventional suppression circuit is an RC network connected directly across the contacts. The capacitor is sized to limit the rate-of-rise of voltage across the gap to below that which initiates an arc. The resistor limits the capacitor discharge current on contact closure; its value is a compromise between maximum rated contact current and limiting the effectiveness of the capacitor. A parallel diode can be added in DC circuits if this compromise cannot be met.

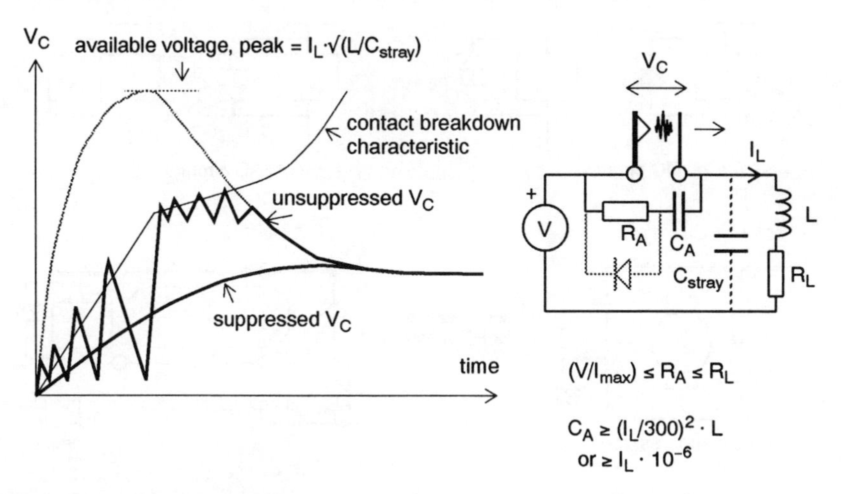

Figure 6.31 Contact noise generation and suppression

6.2.6.1 Suppression of inductive loads

When current through an inductance is interrupted a large transient voltage is generated, governed by $V = -L \cdot di/dt$. Theoretically if di/dt is infinite then the voltage is infinite too; in practice it is limited by stray capacitance if no other measures are taken, and the voltage waveform is a damped sinusoid (if no breakdown occurs) whose frequency is determined by the values of inductance and stray capacitance. Typical examples of switched inductive loads are motors, relay coils and transformers, but even a long cable can have enough distributed inductance to generate a significant transient amplitude. Switching can either be via an electromechanical contact or a semiconductor, and the latter can easily suffer avalanche breakdown due to the overvoltage if the transient is unsuppressed. RF interference is generated in both cases at frequencies determined by stray circuit resonances and is usually radiated from the wiring between switch and load.

The RC snubber circuit can be used in some cases to damp an inductive transient. Other circuits use diode, Zener or varistor clamps as shown in Figure 6.32. Motor interference may appear in common mode with respect to the housing as a result of the high stray capacitance between the housing and the windings, hence you will often need to use common mode decoupling capacitors. In all cases the suppression components must be mounted immediately next to the load terminals, otherwise a radiating current loop is formed by the intervening wiring. Protection of a driver transistor mounted remotely must be considered as a separate function from RF suppression.

6.3 Shielding

Shielding and filtering are complementary practices. There is little point in applying good filtering and circuit design practice to guard against conducted coupling if there is no return path for the filtered currents to take. The shield provides such a return, and also guards against direct field coupling with the internal circuits and conductors.

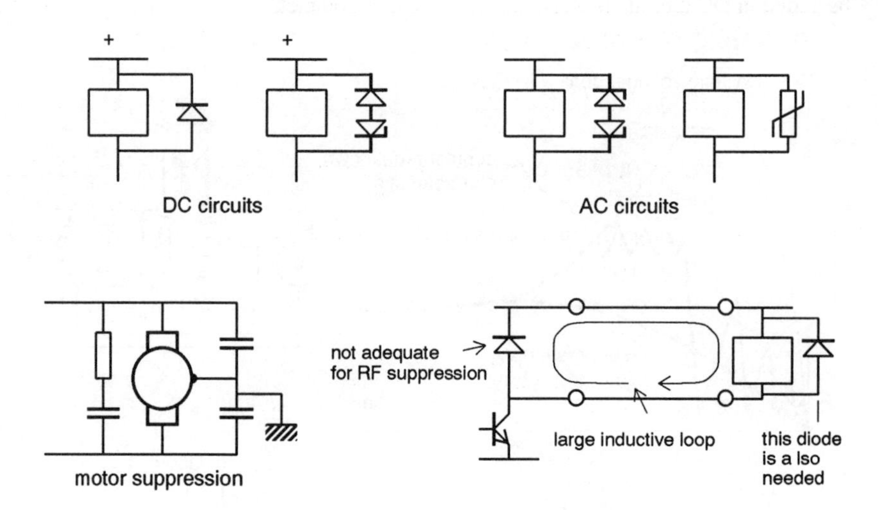

Figure 6.32 Inductive load suppression

Shielding involves placing a conductive surface around the critical parts of the circuit so that the electromagnetic field which couples to it is attenuated by a combination of reflection and absorption. The shield can be an all-metal enclosure if protection down to low frequencies is needed, but if only high frequency (> 30MHz) protection will be enough then a thin conductive coating deposited on plastic is adequate.

Will a shield be necessary?

Shielding is often an expensive and difficult-to-implement design decision, because many other factors – aesthetic, tooling, accessibility – work against it. A decision on whether or not to shield should be taken as early as possible in the project. Chapter 4, sections 4.2 and 4.3 showed that interference coupling is via interface cables and direct induction to/from the pcb. You should be able to calculate to a rough order of magnitude the fields generated by pcb tracks and compare these to the desired emission limit (see section 5.2.2). If the limit is exceeded at this point and the pcb layout cannot be improved, then shielding is essential. Shielding does not of itself affect common mode cable coupling and so if this is expected to be the dominant coupling path a full shield may not be necessary. It does establish a "clean" reference for decoupling common mode currents to, but it is also possible to do this with a large area ground plate if the layout is planned carefully.

6.3.1 Shielding theory

An AC electric field impinging on a conductive wall of infinite extent will induce a current flow in that surface of the wall, which in turn will generate a reflected wave of the opposite sense. This is necessary in order to satisfy the boundary conditions at the wall, where the electric field must approach zero. The reflected wave amplitude determines the reflection loss of the wall. Because shielding walls have finite conductivity, part of this current flow penetrates into the wall and a fraction of it will appear on the opposite side of the wall, where it will generate its own field (Figure

To shield or not to shield

- if predicted differential-mode fields will exceed limits, shielding is essential
- if layout requires dispersed interfaces, shielding will probably be essential
- if layout allows concentrated interfaces, a ground plate may be adequate
- consider shielding only critical circuitry

6.33). The ratio of the impinging to the transmitted fields is one measure of the shielding effectiveness of the wall.

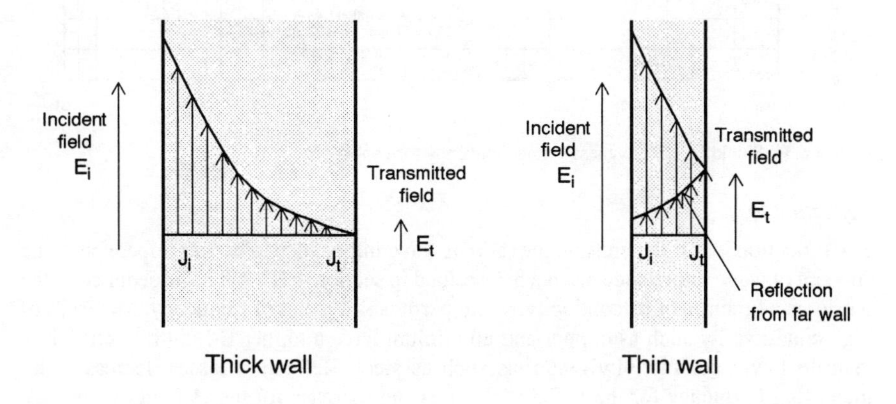

Figure 6.33 Variation of current density with wall thickness

The thicker the wall, the greater the attenuation of the current through it. This absorption loss depends on the number of "skin depths" through the wall. The skin depth (defined in appendix C) is an expression of the electromagnetic property which tends to confine AC current flow to the surface of a conductor, becoming less as frequency, conductivity or permeability increases. Fields are attenuated by 8.6dB (1/e) for each skin depth of penetration. Typically, skin depth in aluminium at 30MHz is 0.015mm.

6.3.1.1 Shielding effectiveness

Shielding effectiveness of a solid conductive barrier describes the ratio between the field strength without the barrier in place, to that when it is present. It can be expressed as the sum of reflection, absorption, and re-reflection losses, as shown in Figure 6.34 and given by equation (6.1):

$$SE(dB) = R(dB) + A(dB) + B(dB) \quad (6.1)$$

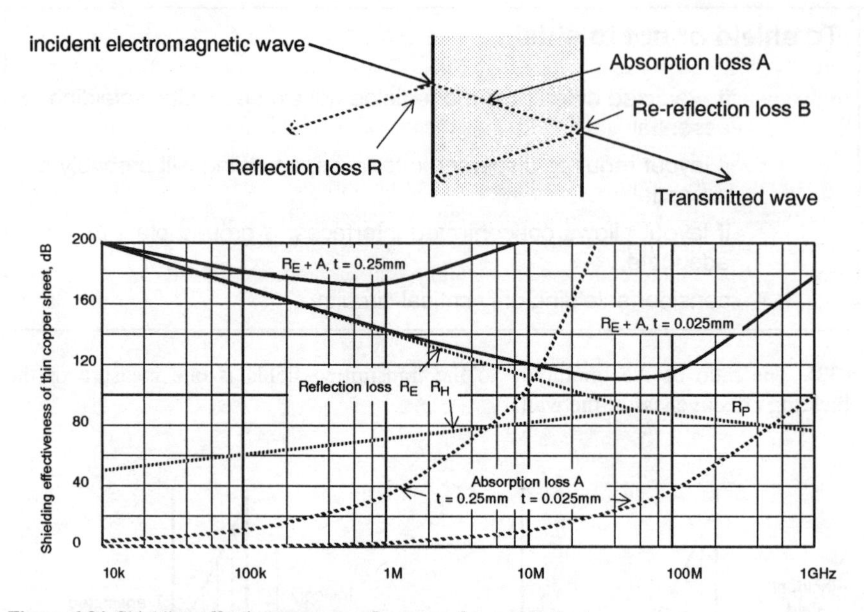

Figure 6.34 Shielding effectiveness versus frequency for copper sheet

Reflection loss

The reflection loss R depends on the ratio of wave impedance to barrier impedance. The concept of wave impedance has been described in section 4.1.3.2. The impedance of the barrier is a function of its conductivity and permeability, and of frequency. Materials of high conductivity such as copper and aluminium have a higher E-field reflection loss than do lower conductivity materials such as steel. Reflection losses decrease with increasing frequency for the E-field (electric) and increase for the H-field (magnetic). In the near field, closer than $l/2\pi$, the distance between source and barrier also affects the reflection loss. Near to the source, the electric field impedance is high and the reflection loss is also correspondingly high. Vice versa, the magnetic field impedance is low and the reflection loss is low. When the barrier is far enough away to be in the far field, the impinging wave is a plane wave, the wave impedance is constant and the distance is immaterial. Refer back to Figure 4.7 on page 101 for an illustration of the distinction between near and far field.

The re-reflection loss B is insignificant in most cases where absorption loss A is greater than 10dB, but becomes important for thin barriers at low frequencies.

Absorption loss

Absorption loss depends on the barrier thickness and its skin depth and is the same whether the field is electric, magnetic or plane wave. The skin depth in turn depends on the barrier material's properties; in contrast to reflection loss, steel offers higher absorption than copper of the same thickness. At high frequencies, as Figure 6.34 shows, it becomes the dominant term, increasing exponentially with the square root of the frequency. Appendix C (section C.4) gives the formulae for the values of A, R and B for given material parameters.

LF magnetic fields

Shielding against magnetic fields at low frequencies is to all intents and purposes impossible with purely conductive materials. This is because the reflection loss to an impinging magnetic field (R_H) depends on the mismatch of the field impedance to the barrier impedance. The low field impedance is well matched to the low barrier impedance and the field is transmitted through the barrier without significant attenuation or absorption. A high-permeability material such as mu-metal or its derivatives can give LF magnetic shielding by concentrating the field within the bulk of the material, but this is a different mechanism to that discussed above, and it is normally only viable for sensitive individual components such as CRTs or transformers.

6.3.2 The effect of apertures

The curves of shielding effectiveness in Figure 6.34 suggest that upwards of 200dB attenuation is easily achievable using reasonable thicknesses of common materials. In fact, the practical shielding effectiveness is not determined by material characteristics but is limited by necessary apertures and discontinuities in the shielding. You will need apertures for ventilation, for control and interface access, and for viewing indicators.

Electromagnetic leakage through an aperture in a thin barrier depends on its longest dimension (d) and the minimum wavelength (λ) of the frequency band to be shielded against. For wavelengths less than or equal to twice the longest aperture dimension there is effectively no shielding. The frequency at which this occurs is the "cut-off frequency" of the aperture. For lower frequencies ($\lambda > 2d$) the shielding effectiveness increases linearly at a rate of 20dB per decade up to the maximum possible for the barrier material, as defined in Figure 6.35.

For most practical purposes shielding effectiveness is determined by the apertures. For frequencies up to 1GHz (the present upper limit for radiated emissions standards) and a minimum shielding of 20dB the maximum hole size allowable is 1.6cm.

6.3.2.1 Windows and ventilation slots

Viewing windows normally involve a large open area in the shield and you have to cover the window with a transparent conductive material, which must make good continuous contact to the surrounding screen, or accept the penalty of shielding at lower frequencies only. You can obtain shielded window components which are laminated with fine blackened copper mesh, or which are coated with an extremely thin film of gold. In either case, there is a tradeoff in viewing quality over a clear window, due to reduced light transmission (between 60 and 80%) and diffraction effects of the mesh. Screening effectiveness of a transparent conductive coating is significantly less than a solid shield, since the coating will have a resistance of a few ohms per square and attenuation will be entirely due to reflection loss. This is not the case with a mesh, but shielding effectiveness of better than 40 – 50dB may be irrelevant anyway because of the effect of other apertures. Shielded windows are also costly and not suited to consumer applications.

Using a sub-enclosure

An alternative method which allows you to retain a clear window, is to shield behind the display with a sub-shield (Figure 6.36), which must of course make good all-round contact with the main panel. The electrical connections to the display must be filtered to preserve the shield's integrity, and the display itself is unshielded and must therefore

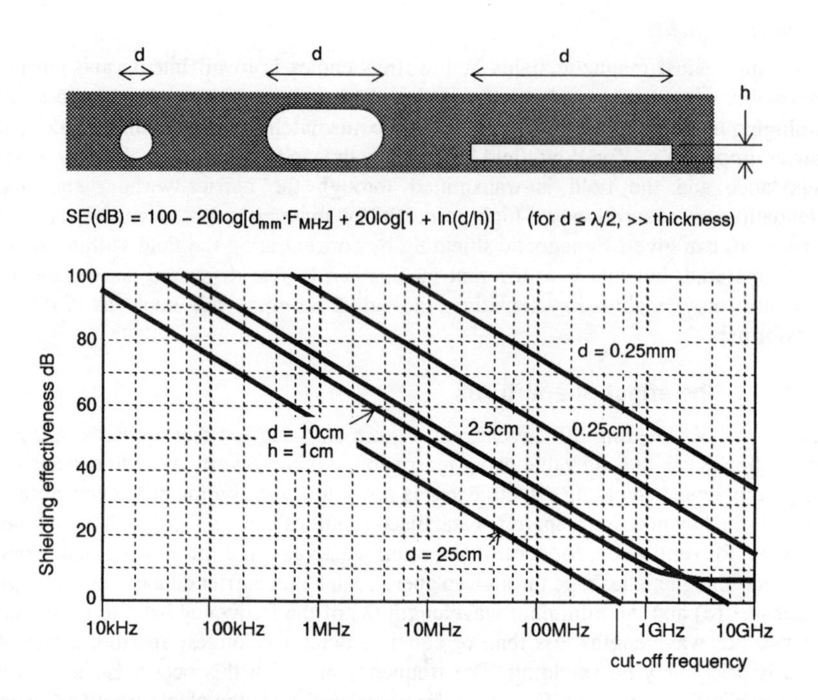

Figure 6.35 Shielding effectiveness degradation due to apertures

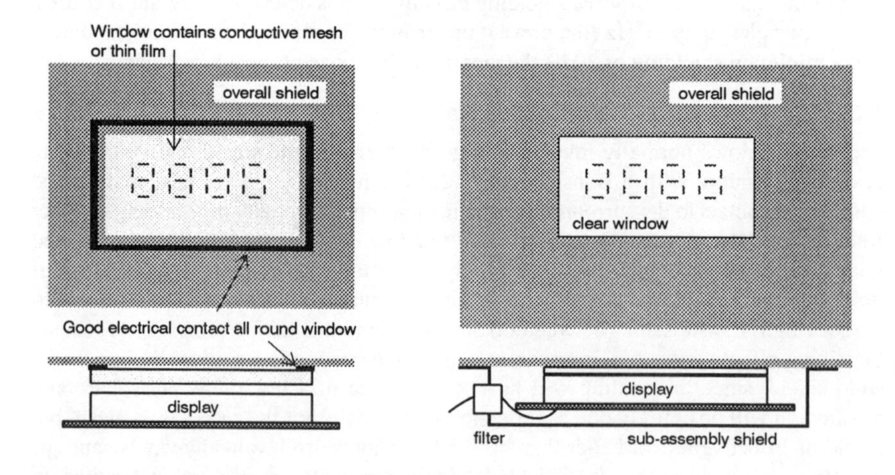

Figure 6.36 Alternative ways to shield a display window

not be susceptible nor contain emitting sources. This alternative is frequently easier and cheaper than shielded windows.

Mesh and honeycomb

Ventilation holes can be covered with a perforated mesh screen, or the conductive panel may itself be perforated. If individual equally sized perforations are spaced close together (hole spacing < $\lambda/2$) then the reduction in shielding over a single hole is approximately proportional to the square root of the number of holes. Thus a mesh of 100 4mm holes would have a shielding effectiveness 20dB worse than a single 4mm hole. Two similar apertures spaced greater than a half-wavelength apart do not suffer any significant extra shielding reduction.

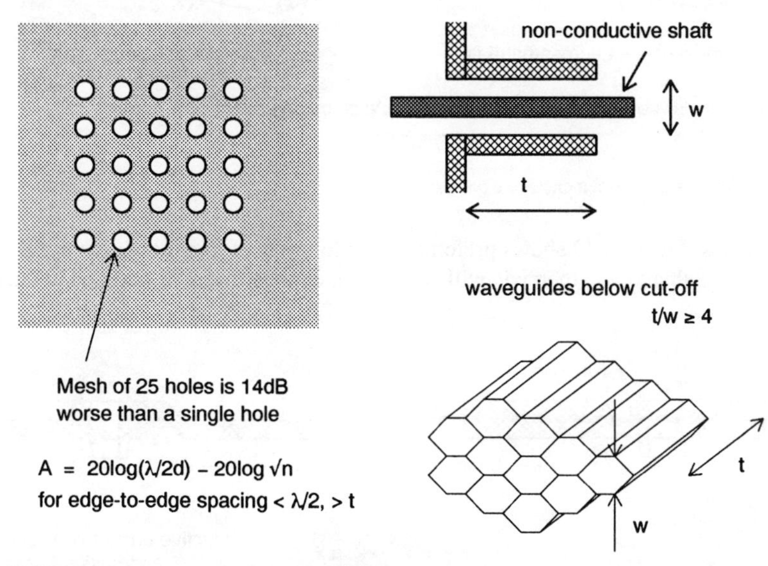

Figure 6.37 Mesh panels and the waveguide below cut-off

You can if necessary gain improved shielding of vents, at the expense of thickness and weight, by using "honeycomb" panels in which the honeycomb pattern functions as a waveguide below cut-off (Figure 6.37). In this technique the shield thickness is several times that of the width of each individual aperture. A common t/w ratio is 4:1 which offers an intrinsic shielding effectiveness of over 100dB. This method can also be used to conduct insulated control spindles (*not* conductive ones!) through a panel.

6.3.2.2 The effect of seams

An electromagnetic shield is normally made from several panels joined together at seams. Unfortunately, when you join two sheets the electrical conductivity across the joint is imperfect. This may be because of distortion, so that surfaces do not mate perfectly, or because of painting, anodising or corrosion, so that an insulating layer is present on one or both metal surfaces.

Consequently, the shielding effectiveness is reduced by seams almost as much as it is by apertures (Figure 6.38). The ratio of the fastener spacing d to the seam gap h is high enough to improve the shielding over that of a large aperture by 10 – 20dB. The problem is especially serious for hinged front panels, doors and removable hatches that form part of a screened enclosure. It is mitigated to some extent if the conductive sheets overlap, since this forms a capacitor which provides a partial current path at high

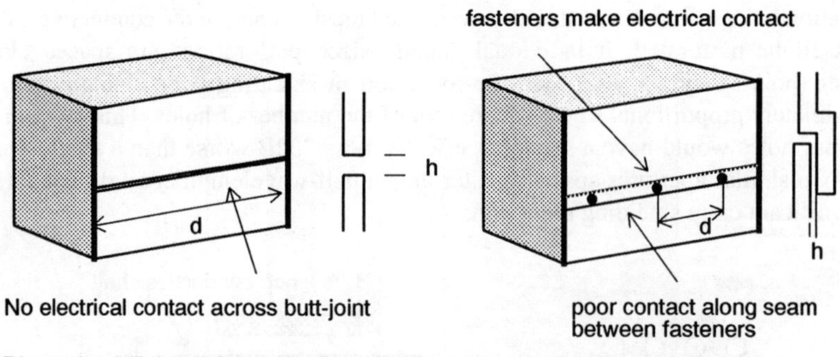

Figure 6.38 Seams between enclosure panels

frequencies. Figure 6.39 shows preferred ways to improve joint conductivity. If these are not available to you then you will need to use exra hardware as outlined in the next section.

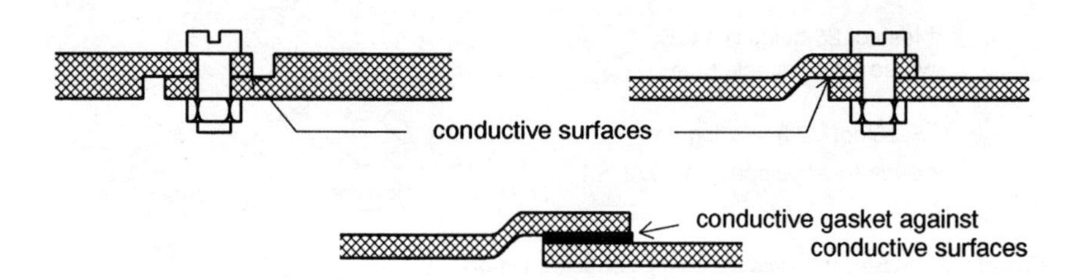

Figure 6.39 Cross sections of joints for good conductivity

6.3.2.3 Seam and aperture orientation

The effect of a joint discontinuity is to force shield current to flow around the discontinuity. If the current flowing in the shield were undisturbed then the field within the shielded area would be minimized, but as the current is diverted so a localised discontinuity occurs, and this creates a field coupling path through the shield. The shielding effectiveness graph shown in Figure 6.35 assumes a worst case orientation of current flow. A long aperture or narrow seam will have a greater effect on current flowing at right angles to it than on parallel current flow†. This effect can be exploited if you can control the orientation of susceptible or emissive conductors within the shielded environment (Figure 6.40).

The practical implication of this is that if all critical internal conductors are within the same plane, such as on a pcb, then long apertures and seams in the shield should be aligned parallel to this plane rather than perpendicular to it. Generally, you are likely to obtain an advantage of no more than 10dB by this trick, since the geometry of internal

† Antenna designers will recognise that this describes a slot antenna, the reciprocal of a dipole.

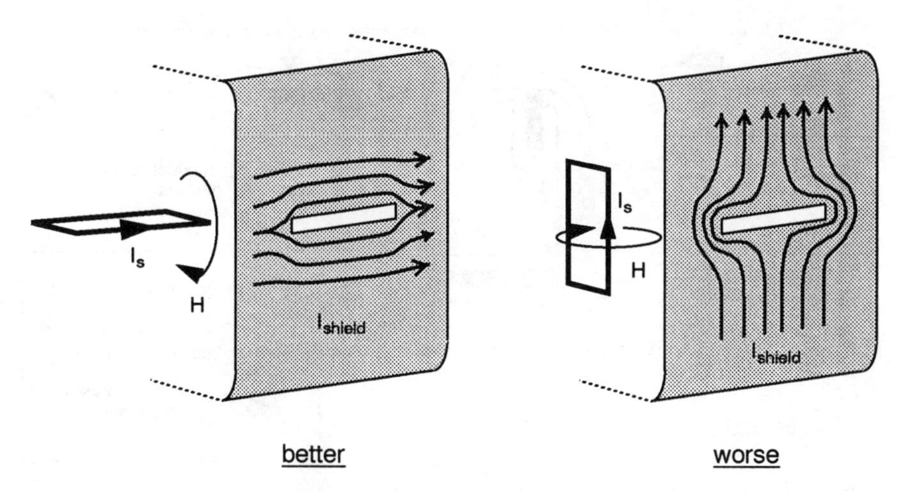

Figure 6.40 Current loop versus aperture orientation

conductors is never exactly planar. Similarly cables or wires, if they must be routed near to the shield, should be run along or parallel to apertures rather than across them. But because the leakage field coupling due to joint discontinuities is large near the discontinuity, internal cables should preferably not be routed near to apertures or seams.

Apertures on different surfaces or seams at different orientations can be treated separately since they radiate in different directions.

6.3.3 Shielding hardware

Many manufacturers offer various materials for improving the conductivity of joints in conductive panels. From the advertising hype, you might think that these are all you need to rid yourself of EMC problems for good. In fact such materials can be useful if properly applied, but they must be used with an awareness of the principles discussed above, and their expense will often rule them out for cost-sensitive applications except as a last resort.

6.3.3.1 Gaskets and contact strip

Shielding effectiveness can be improved by reducing the spacing of fasteners between different panels. If you need effectiveness to the upper limit of 1GHz, then the necessary spacing becomes unrealistically small when you consider maintenance and accessibility. In these cases the conductive path between two panels or flanges can be improved by using any of the several brands of conductive gasket, knitted wire mesh or finger strip that are available. The purpose of these components is to be sandwiched in between the mating surfaces to ensure continuous contact across the joint, so that shield current is not diverted (Figure 6.41). Their effectiveness depends entirely on how well they can match the impedance of the joint to that of the bulk shield material.

You must bear in mind a number of factors when selecting a conductive gasket or finger material:

- its conductivity: it should be of the same order of magnitude as the panel

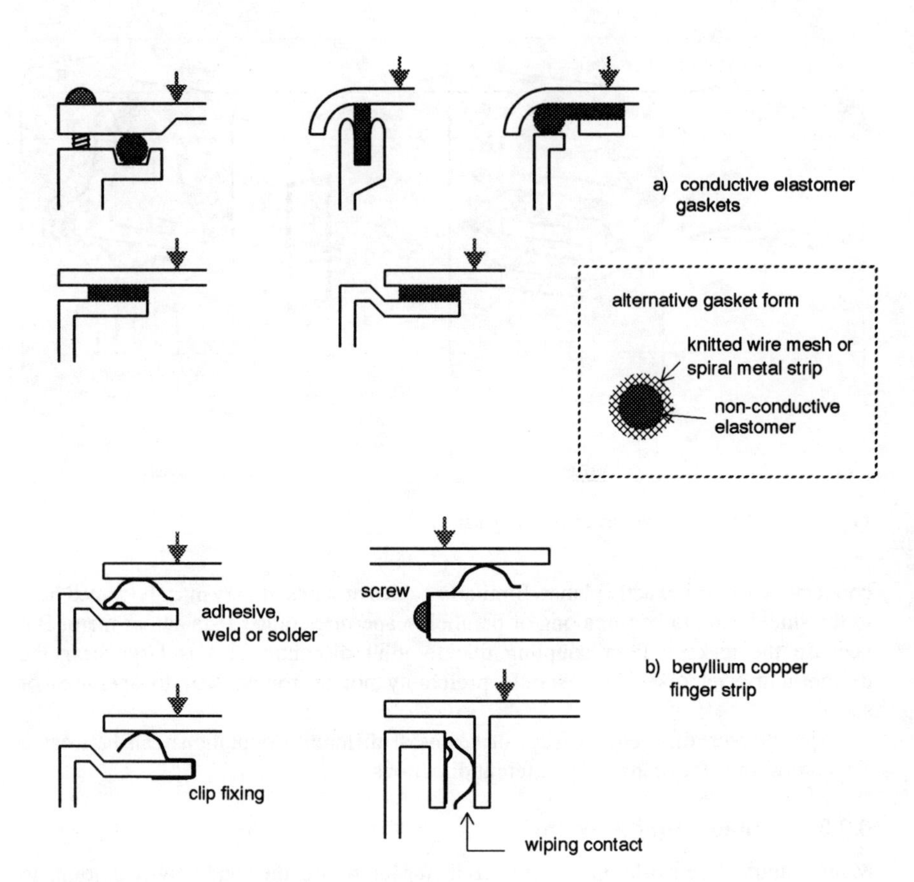

Figure 6.41 Usage of gaskets and finger strip

material;

- ease of mounting: gaskets should normally be mounted in channels machined or cast in the housing, and the correct dimensioning of these channels is important to maintain an adequate contact pressure without over-tightening. Finger strip can be mounted by adhesive tape, welding, riveting, soldering, fasteners or by just clipping in place. The right method depends on the direction of contact pressure;
- galvanic compatibility with its host: to reduce corrosion, the gasket metal and its housing should be close together and preferably of the same group within the electrochemical series (Table 6.3). The housing material should be conductively finished – alochrome or alodine for aluminium, nickel or tin plate for steel;
- environmental performance: conductive elastomers will offer combined electrical and environmental protection, but may be affected by moisture, fungus, weathering or heat. If you choose to use separate environmental and conductive gaskets, the conductive gasket should be placed *inside* the environmental seal, and also inside screw mounting holes.

Anodic - most easily corroded →				
Group I	Group II	Group III	Group IV	GroupV
Magnesium	Aluminium + alloys			
	Zinc	Carbon steel		
	Chromium	Iron	Nickel	
	Galvanized iron	Cadmium	Tin, solder	Copper + alloys
			Lead	Silver
			Brass	Palladium
			Stainless steel	Platinum
				Gold
			→ Cathodic - least easily corroded	
Corrosion occurs when ions move from the more anodic metal to the more cathodic, facilitated by an electrolytic transport medium such as moisture or salts				

Table 6.3 The electrochemical series

6.3.3.2 Conductive coatings

Many electronic products are enclosed in plastic cases for aesthetic or cost reasons. These can be made to provide a degree of electromagnetic shielding by covering one or both sides with a conductive coating [23][72]. Normally, this involves both a moulding supplier and a coating supplier. You can also use conductively filled plastic composites to obtain a marginal degree of shielding (around 20dB); it is debatable whether the extra material cost justifies such an approach, considering that better shielding performance can be offered by conductive coating at lower overall cost [25]. Conductive fillers affect the mechanical and asthetic properties of the plastic, but their major advantage is that no further treatment of the moulded part is needed. Another problem is that the moulding process may leave a "resin rich" surface which is not conductive, so that the conductivity across seams and joints is not assured. As a further alternative, metallized fabrics are now becoming available which can be incorporated into some designs of compression moulding.

Shielding performance

The same dimensional considerations apply to apertures and seams as for metal shields. Thin coatings will be almost as effective against electric fields at high frequencies as solid metal cases but are ineffective against magnetic fields. The major shielding mechanism is reflection loss (Figure 6.34) since absorption is negligible except at very high frequencies, and re-reflection (B) will tend to reduce the overall reflection losses. The higher the resistivity of the coating, the less its efficiency. For this reason conductive paints, which have a resistivity of around 1Ω/square, are poorer shields than the various types of metallization (see Table 6.4) which offer resistivities below 0.1Ω/square.

Enclosure design

Resistivity will depend on the thickness of the coating, which in turn is affected by factors such as the shape and sharpness of the moulding – coatings will adhere less to, and abrade more easily from, sharp edges and corners than rounded ones. Ribs, dividing walls and other mould features that exist inside most enclosures make application of sprayed-on coatings (such as conductive paint or zinc arc spray) harder and favour the electroless plating methods. Where coatings must cover such features, your moulding design should include generous radii, no sharp corners, adequate space between ribs and no deep or narrow crevices.

Coating properties

Environmental factors, particularly abrasion resistance and adhesion, are critical in the selection of the correct coating. Major quality considerations are

- will the coating peel or flake off into the electrical circuitry?
- will the shielding effectiveness be consistent from part to part?
- will the coating maintain its shielding effectiveness over the life of the product?

Adhesion is a function of thermal or mechanical stresses, and is checked by a non-destructive tape or destructive cross-hatch test. Typically, the removal of any coating as an immediate result of tape application constitutes a test failure. During and at completion of thermal cycling or humidity testing, small flakes at the lattice edges after a cross-hatch test should not exceed 15% of the total coating removal.

Electrical properties should not change after repeated temperature/humidity cycling within the parameters agreed with the moulding and coating suppliers. Resistance measurements should be taken from the farthest distances of the test piece and also on surfaces critical to the shielding performance, especially mating and grounding areas.

Table 6.4 compares the features of the more commonly available conductive coatings (others are possible but are more expensive and little used). These will give

	Cost £/m^2	E-field shielding	Thickness	Adhesion	Scratch resistance	Maskable	Comments
Conductive Paint (nickel, copper)	5 - 15	Poor/ average	0.05mm	Poor	Poor	Yes	Suitable for prototyping
Zinc Arc Spray (zinc)	5 - 10	Average/ good	.1-.15mm	Depends on surface prep	Good	Yes	Rough surface, inconsistent
Electroless Plate (copper, nickel)	10 - 15	Average/ good	1-2μm	Good	Poor	No	Cheaper if entire part plated
Vacuum Metallization (aluminium)	10 - 15	Average	2-5μm	Depends on surface prep	Poor	Yes	Poor environ-mental qualities

Table 6.4 Comparison of conductive coating techniques

shielding effectiveness in the range of 30 – 70dB if properly applied. It is difficult to compare the shielding effectiveness figures from different manufacturers unless they specify very clearly the methods used to perform their shielding effectiveness tests; different methods do not give comparable results. Also, laboratory test methods do not necessarily correlate with the performance of a practical enclosure for a commercial product.

Appendices

Appendix A

Design checklist

Many factors must be considered when looking at the EMC aspects of a design, and it is easy to overlook an important point. This checklist is provided for you to assess your design against as it proceeds.

- Design for EMC from the beginning; know what performance you require
- Partition the system into critical and non-critical sections
  - determine which circuits will be noisy or susceptible and which will not
  - lay them out in separate areas as far as possible
  - select internal and external interface points to allow optimum common-mode current control
- Select components and circuits with EMC in mind:
  - use slow and/or high-immunity logic
  - use good rf decoupling
  - minimise signal bandwidths, maximise levels
  - provide power supplies of adequate (noise-free) quality
  - incorporate a watchdog circuit on every microprocessor
- PCB layout:
  - ensure proper signal returns; if necessary include isolation to define preferred current paths
  - keep interference paths segregated from sensitive circuits
  - minimise ground inductance with thick gridding or ground plane
  - minimise enclosed loop areas in high current, high di/dt or sensitive circuits
  - minimise surface areas of nodes with high dv/dt
  - minimise track and component leadout lengths
- Cables:
  - avoid parallel runs of signal and power cables
  - use signal cables and connectors with adequate screening
  - use twisted pair if appropriate
  - run cables away from apertures in the shielding, close to conductive grounded structures
  - avoid resonant lengths where possible, consider damping cables with ferrite suppressors
  - ensure that cable screens are properly terminated to the correct type of connector
- Grounding:
  - design and enforce the ground system at the product definition stage
  - consider the ground system as a return current path, not as 0V reference
  - ensure adequate bonding of screens, connectors, filters, cabinets etc
  - ensure that bonding methods will not deteriorate in adverse environments

- mask paint from any intended conductive areas
- keep earth leads short and define their geometry
- avoid common ground impedances
- provide a "clean" ground area for decoupling all interfaces

- Filters:
  - optimise the mains filter for the application
  - use correct components and filter configuration for I/O lines
  - ensure a good ground return for each filter
  - apply filtering to interference sources, such as switches or motors
  - pay attention to the layout of filter components and associated wiring or tracks
- Shielding:
  - determine the type and extent of shielding required from the frequency range of interest
  - enclose particularly sensitive or noisy areas with extra internal shielding
  - avoid large or resonant apertures in the shield, or take measures to mitigate them
  - ensure that separate panels are well bonded along their seams
- Test and evaluate for EMC continuously as the design progresses

Appendix B

EMC test and control plans

B.1 Control Plan

B.1.1 The purpose of the control plan

The EMC control plan is a document, part of the specification of a new product, which lays down a schedule and a method to define how work towards the product's EMC compliance will be undertaken. Some projects within the military and aerospace sectors require this plan to be submitted as part of the tender documentation. Even if it is not a contractual requirement, the use within the design team of a detailed plan, showing which actions need to be taken and when, is a discipline that will reap benefits at the tail end of the design process when EMC performance comes to be evaluated. Too often if this discipline has not been applied the EMC performance turns out to be the Achilles heel of the product. Within a properly structured design environment the incorporation of an EMC control plan is not a major overhead.

Responsibility for EMC can be vested in one individual who is a member of the product design team. Their task is to develop the EMC control and test plans and to define the performance and test failure criteria for the product. Familiarity with the appropriate standards and test methods as they apply to the product is essential. Structuring the EMC control in this way results in a strong "sense of ownership" of the EMC aspects of the product. Alternatively, the company may have established a separate EMC control group, who oversee the EMC aspects of all product developments on an internal consultancy basis. Whilst this approach allows a company to build up a strong core of EMC-specific expertise, it may result in friction between the product development team and the EMC control engineers which may not eventually result in the optimum product-specific solution.

A properly structured and documented control plan will also form a valuable and major part of the Technical Construction File (section 1.3.2) if this route to compliance is being followed. If you are taking the self certification route, it will provide powerful evidence if your Declaration of Conformity is ever challenged.

B.1.2 Contents

The control plan can be divided into two major sections, one defining best design practice which is to be followed in developing the product, the other defining project reviews and control stages.

1 Definition of EMC phenomena to be addressed

2 Design practice

This section can draw on the various EMC control methods as discussed in chapters 5 and 6 of this book. Recommendations should be as detailed as possible given what is already known about the design of the product.

- Grounding regime, including a ground map
- Control and layout of interfaces
- Use of screened connectors and cable
- PCB layout
- Circuit techniques
- Choice of power supply
- Filtering of power ports
- Filtering and isolation of signal ports
- Packaging design including screening

3 Project management

Mandatory EMC design reviews, control stages and checkpoints, and who is responsible for overseeing them; these can normally be incorporated into the overall project management scheme but there may be advantages in separating out the EMC functions

- Preliminary design review
- Design testing
- Detailed design review
- Pre-compliance confidence testing
- Final design review
- Compliance test or certification
- Responsibility for Declaration of Conformity
- Production quality assurance

B.2 Test Plan

B.2.1 The need for a test plan

An EMC test plan is a vital part of the specification of a new product. It provides the basis around which the development and pre-production stages can be structured in order for the product to achieve EMC compliance with the minimum of effort. It can act as the schedule for in-house testing, or can be used as a contractual document in dealing with an external test house. Although you should prepare it as soon as the project gets underway, it will of necessity need revision as the product itself develops and especially in the light of actual test experience on prototype versions of the product.

B.2.2 Contents

1 Description of the equipment under test (EUT)

The basic description of the EUT must specify the model number and which (if any) variants are to be tested under this generic model type.

- is EUT stand-alone or part of a larger system?

If it is to be tested as a stand-alone unit then no further information is needed. If it can only be tested as part of a system, e.g. it is a plug-in module or a computer peripheral, then the components of the system of which it is a part must also be specified. You need to take care that the test results will not be compromised by a failure on the part of other system components.

- system configuration and criteria for choosing it

  Following on from the above, if the EUT can form part of a system or installation which may contain many other different components, you will need to specify a representative system configuration which will allow you to perform the tests. The criteria on which the choice of configuration is based, i.e. how you decide what is "representative", must be made clear.

- revision level and acceptable revision changes before/during testing

  During design and development you will want to perform some confidence tests. At this stage the build state must be carefully defined, by reference to drawing status, even if revision levels have not yet been issued. Once the equipment reaches the stage of qualification approval the tests must be done on a specimen which is certified as being built to a specific revision level, which must be the same as that which is placed on the market.

2 Statement of test objectives

Self-evidently, you will need to state why the tests are to be performed.

- to meet legal standards (EMC Directive, FCC certification)

  As will be necessary to be legally permitted to place the equipment on the market

- to meet voluntary standards

  To improve your competitive advantage

- to meet a contractual obligation

  Because EMC performance has been written in to the procurement contract for the equipment or system

3 The tests to be performed:

- frequency ranges and voltage levels to be covered

  These are normally specified in the standard(s) you have chosen. If you are not using standards, or are extending or reducing the tests, this must be made clear.

- test equipment and facility to be used

  This will also be determined by the standard(s) in use. Some standards have specific requirements for test equipment, e.g. CISPR16 instrumentation. If you will be using an external test house they will determine the instrumentation that they will need to use to cover the required tests. If you are doing it in house, it is your responsibility.

- location of test points

  The number of lines (or ports – a definition which includes the "enclosure port") to be tested directly influences the test time. The standard(s) you choose to apply may define which lines should be tested. In some cases you can test just one representative line and claim that it covers all others of the same type. The position of the test point can be critical, especially for electrostatic discharge application, and must be specified.

- EUT operating modes

  If there are several different operating modes then you may be able to identify a worst case mode which includes the majority of operating scenarios and emission/ susceptibility profiles. This will probably need some exploratory testing. It has a direct influence on the testing time. The rate at which a disturbance is applied or an emission measurement is made also depends on the cycle time of the specified operating mode.

- test schedule, including sequence of tests and EUT operating modes

  The order in which tests are applied and the sequence of operating modes may or may not be critical, but should be specified.

4 Criteria for determining locations of monitoring or injection points

It is important that both you and any subsequent assessment authority know why you have chosen to apply tests to particular points on the EUT. These are often specified in the chosen standard(s), for instance the mains lead for conducted emissions. But e.g. the choice of ESD application points should be supported by an assessment of likely use of the equipment and/or some preliminary testing to determine weak points.

5 Description of EUT exercising software and any necessary ancillary equipment or simulators

The EUT may benefit from special software to fully exercise its operating modes; if it is not stand-alone it will need some ancillary support equipment. Both of these should be calibrated or declared fit for purpose. If the support equipment will be housed in a separate chamber it can be interfaced via a filtered bulkhead which will reduce fortuitous emissions and isolate it from disturbances applied to the EUT. This filtering arrangement will need to be specified.

6 Requirements of the test facility, including:

- environmental conditions

  Special requirements for temperature, humidity, vibration etc.
- safety precautions needed

  E.g. if the EUT uses ionising radiations or extra high voltages, is dangerously heavy or hot, or if the tests require high values of radiated field
- special handling and functional support equipment

  Fork lift trucks, large turntables, hydraulic or air/water cooling services etc.
- power sources

  AC or DC voltage, current, frequency, single- or three-phase, VA rating and surge current requirement, total number of power lines (remember that FCC certification testing will need US-standard power supplies)
- system command software

  Will the tests require special software to be written to integrate the test suite with the EUT operation?
- security classification

7 Sketch and details of test set-up, including:

- physical location and layout of EUT and test instrumentation

  This is defined in general terms in the various standards, but you will normally have to interpret the instructions given in these to apply to your particular EUT. Critical points are distances, orientation and proximity to other objects, especially the ground plane. The final test report should include photographs which record the set-up.
- electrical interconnections

  Cable layout and routing has a critical effect at high frequencies and must be closely defined. You should also specify types of connector and cable to the EUT, if they would otherwise go by default.

8 Type of report to be issued

This can range from a test house certificate which merely states whether the EUT did or did not meet its specification, to a detailed test report which includes all results and test procedures. Most UK test houses would deliver a report to NAMAS standards which contains essential results and information without detailing lab procedures. For the purpose of complying with the EMC Directive, you need to know whether you want a certificate of compliance with an EN standard or a report for insertion into a technical file.

9 How to evaluate test results

- computation of safety margins

  A major part of the test results will be the levels of EUT emissions, or the levels of disturbance at which susceptibility effects occur. There needs to be an accepted way of deriving safety margins between these levels and specification limits, determined by known measurement uncertainties and the operational parameters of the system in which the EUT is situated. These margins are essential to an interpretation of the test results.

- acceptance criteria

  See B.2.4.

- how to deal with ambient signals

  In emissions testing, those signals which are already present in the test environment may be of a sufficiently high level to mask the EUT emissions. Some emissions standards define the procedure to be followed if this is the case.

B.2.3 Test and calibration procedures

Once the test plan has been written, the procedures which the EMC test technician is to follow can be derived from it. Alternatively, the test procedures may be written as an integral part of the test plan itself. Necessary ingredients of the test procedure specification are:

- calibration of test instrumentation

  At a NAMAS accredited test house, all items of test equipment are regularly calibrated. For in-house testing you may need to establish separate calibration procedures.

- description of automated testing procedures

  Typically, much of the measurement procedure will be handled by software which controls functions such as frequency ranges and sweep rates, field strength levelling, transducer selection, bandwidths and detector function, and may also operate antenna and EUT positioning apparatus such as turntables and towers. If this software is not already well established you will need to specify it in detail.

- description of manual testing procedures

  For those parts of the testing which cannot be automated, such as hand application of ESD transients, and for the setting up requirements of each test.

- test schedule

  A list of the tests to be applied, the operating mode(s) of the EUT in each case and the order in which to do them.

B.2.4 Susceptibility criteria

When you perform immunity testing, it is essential to be able to judge whether the EUT has in fact passed or failed the test. This in turn demands a statement of the minimum acceptable performance which the EUT must maintain during and after testing. Such a statement can only be related to the EUT's own functional operating specification. In order to facilitate preparing this statement, the generic immunity standards [94] contain a set of guidelines for the criteria against which the operation of the EUT can be judged, and which are used to formulate the acceptance criteria for a given EUT against specific tests:

> Performance criterion A: The apparatus shall continue to operate as intended. No degradation of performance or loss of function is allowed below a performance level specified by the manufacturer, when the apparatus is used as

intended. In some cases the performance level may be replaced by a permissible loss of performance. If the minimum performance level or the permissible performance loss is not specified by the manufacturer then either of these may be derived from the product description and documentation (including leaflets and advertising) and what the user may reasonably expect from the apparatus if used as intended.

Performance criterion B: The apparatus shall continue to operate as intended after the test. No degradation of performance or loss of function is allowed below a performance level specified by the manufacturer, when the apparatus is used as intended. During the test, degradation of performance is however allowed. No change of actual operating state or stored data is allowed. If the minimum performance level or the permissible performance loss is not specified by the manufacturer then either of these may be derived from the product description and documentation (including leaflets and advertising) and what the user may reasonably expect from the apparatus if used as intended.

Performance criterion C: Temporary loss of function is allowed, provided the loss of function is self recoverable or can be restored by the operation of the controls.

Appendix C

Useful tables and formulae

C.1 The deciBel

The deciBel (dB) represents a logarithmic ratio between two quantities. Of itself it is unitless. If the ratio is referred to a specific quantity (P_2, V_2 or I_2 below) this is indicated by a suffix, e.g. dBμV is referred to 1μV, dBm is referred to 1mW.

Common suffixes

suffix	refers to	suffix	refers to
dBV	1 volt	dBA	1 amp
dBmV	1 millivolt	dBμA	1 microamp
dBμV	1 microvolt	dBμA/m	1 microamp per metre
dBV/m	1 volt per metre	dBW	1 watt
dBμV/m	1 microvolt per metre	dBm	1 milliwatt
		dBμW	1 microwatt

Originally the dB was conceived as a power ratio, hence it is given by

$$\text{dB} = 10 \log_{10} (P_1/P_2)$$

Power is proportional to voltage squared, hence the ratio of voltages or currents across a constant impedance is given by

$$\text{dB} = 20 \log_{10} (V_1/V_2) \quad \text{or} \quad 20 \log_{10} (I_1/I_2)$$

Conversion between voltage in dBμV and power in dBm for a given impedance Z ohms is

$$V(\text{dB}\mu\text{V}) = 90 + 10 \log_{10} (Z) + P(\text{dBm})$$

dBμV versus dBm for Z = 50Ω

dBμV	μV	dBm	pW	dBμV	mV	dBm	nW
–20	0.1	–127	0.0002	30	0.03162	–77	0.02
–10	0.316	–117	0.002	40	0.10	–67	0.2
				50	0.3162	–57	2.0
0	**0**	**–107**	**0.02**	60	1.0	–47	20.0
							μW
5	1.778	–102	0.063	70	3.162	–37	0.2
7	2.239	–100	0.1	80	10.0	–27	2.0
10	3.162	–97	0.2	90	31.62	–17	20.0
15	5.623	–92	0.632	100	100.0	–7	200.0
20	10.0	–87	2.0	**120**	**1.0V**	**+13**	**20mW**

Actual voltage, current or power can be derived from the antilog of the dB value:

$$V = \log^{-1}(\text{dBV}/20) \text{ volts}$$
$$I = \log^{-1}(\text{dBA}/20) \text{ amps}$$
$$P = \log^{-1}(\text{dBW}/10) \text{ watts}$$

Table of ratios

dB	Voltage or current ratio	Power ratio	dB	Voltage or current ratio	Power ratio
−30	0.0316	0.001	12	3.981	15.849
−20	0.1	0.01	14	5.012	25.120
−10	0.3162	0.1	16	6.310	39.811
−6	0.501	0.251	18	7.943	63.096
−3	0.708	0.501	20	10.000	100.00
0	**1.000**	**1.000**	25	17.783	316.2
1	1.122	1.259	30	31.62	1000
2	1.259	1.585	35	56.23	3162
3	1.413	1.995	40	100.0	10,000
4	1.585	2.512	45	177.8	31,623
5	1.778	3.162	50	316.2	10^5
6	1.995	3.981	60	1000	10^6
7	2.239	5.012	70	3162	10^7
8	2.512	6.310	80	10,000	10^8
9	2.818	7.943	90	31,623	10^9
10	3.162	10.000	100	10^5	10^{10}
			120	10^6	10^{12}

C.2 Antennas

Antenna factor

$$AF = E - V$$

where AF = antenna factor, dB/m
E = field strength at the antenna, dBμV/m
V = voltage at antenna terminals, dBμV

Gain versus antenna factor

$$G = 20 \log F - 29.79 - AF$$

where G = gain over isotropic antenna, dBi
F = frequency, MHz
AF = antenna factor, dB/m

Dipoles

Gain of a $\lambda/2$ dipole over an isotropic radiator

$$G = 1.64 \text{ or } 2.15\text{dB}$$

Input resistance of short dipoles of length L [16]

$0 < L < \lambda/4$: $R_{in} = 20 \cdot \pi^2 \cdot (L/\lambda)^2$ ohms

$\lambda/4 < L < \lambda/2$: $R_{in} = 24.7 \cdot (\pi \cdot L/\lambda)^{2.4}$ ohms

VSWR

The term Voltage Standing Wave Ratio (VSWR) describes the degree of mismatch between a transmission line and its source or load. It also describes the amplitude of the standing wave that exists along the line as a result of the mismatch.

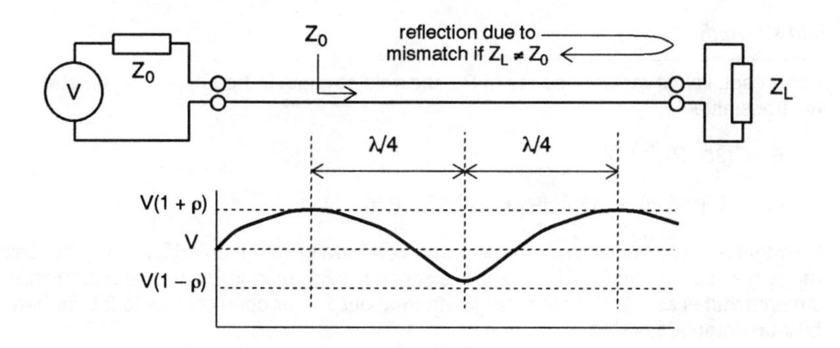

VSWR $K = (1 + \rho)/(1 - \rho) = (Z_0/Z_L)$ or (Z_L/Z_0)

Reflection coefficient $\rho = (K - 1)/(K + 1)$

C.3 Fields

The wave impedance

In free space,

$$Z_0 = (\mu_0/\varepsilon_0)^{0.5} = E/H = 377\Omega \text{ or } 120\pi$$

μ_0 = permeability of free space = $4\pi \cdot 10^{-7}$ Henries per metre
ε_0 = permittivity of free space = $8.85 \cdot 10^{-12}$ Farads per metre

Near field / far field

$d < \lambda/2\pi$: near field; $d > \lambda/2\pi$: far field

(see Figure 4.7 on page 101)

In the near field, the impedance is either higher or lower than Z_0 depending on its source. For a high impedance field of F Hz at distance d metres due to an electric dipole,

$$|Z| = 1/(2\pi F \cdot \varepsilon \cdot d)$$

For a low impedance field due to a current loop,

$$|Z| = 2\pi F \cdot \mu \cdot d$$

Power density

Conversion from field strength to power density in the far field

$$P = E^2/1200\pi$$

where P = power density, mW/cm^2
E is field strength, volts/metre

or for an isotropic antenna

$$P = P_T/4\pi \cdot R^2$$

where R is distance in metres from source of power P_T watts

Field strength

For an equivalent radiated power of P_T, the field strength in free space at R metres from the transmitter is

$$E = (30 \cdot P_T)^{0.5} / R$$

or $$E\ (\text{mV/m}) = 173 \cdot (P_T \text{ in kW})^{0.5} / (R \text{ in km})$$

Propagation near the ground is attenuated at a greater rate than 1/R. For the frequency range between 30 and 300MHz and distances greater than 30 metres, the median field strength varies as $1/R^n$ where n ranges from about 1.3 for open country to 2.8 for heavily built-up urban areas [86].

Field strength from a small loop or monopole [10]

For a loop in free space of area A m^2 carrying current I amps at a frequency f Hz, electric field at distance R metres is

$$E = 131.6 \cdot 10^{-16} (f^2 \cdot A \cdot I)/R \cdot \sin\theta \quad \text{volts/metre}$$

Correcting for ground reflection (x2) with a measuring distance of 10 metres at maximum orientation,

$$E = 26.3 \cdot 10^{-16} (f^2 \cdot A \cdot I) \quad \text{volts/metre}$$

Short monopole of length L ($<< \lambda/2$) over ground plane at distance R driven by common mode current I

$$E = 4\pi \cdot 10^{-7} \cdot (f \cdot I \cdot L)/R \cdot \sin\theta \quad \text{volts/metre}$$

Maximum orientation at 10 metres,

$$E = 1.26 \cdot 10^{-7} \cdot (f \cdot I \cdot L) \quad \text{volts/metre}$$

Electric versus magnetic field strength

In the far field the electric and magnetic field strengths are related by the impedance of free space, Z_o (377Ω).

$$E(\text{dB}\mu\text{V/m}) = H(\text{dB}\mu\text{A/m}) + 51.5$$

H can be expressed in Amps per metre, Tesla or Gauss

$$1 \text{ Gauss} = 100\mu\text{T} = 79.5 \text{ A/m}$$

$$1 \text{ A/m} = 4\pi \cdot 10^{-7} \text{ T}$$

The field equations [7]

The following equations characterize the E and H fields at a point P due to an elementary electric dipole (current filament) and an elementary magnetic dipole (current loop). They use the spherical co-ordinate system shown below.

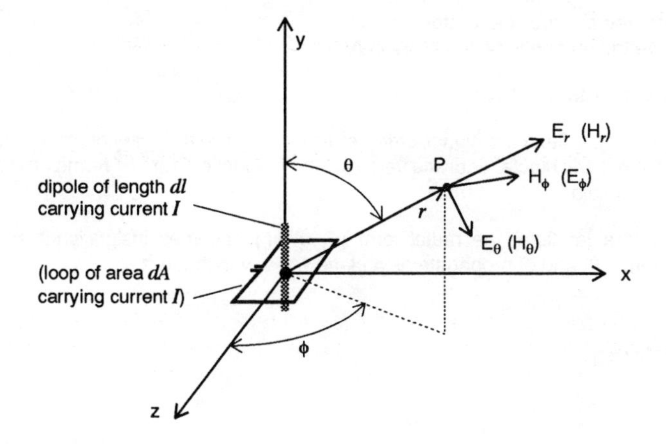

For the electric dipole,

$$E_r = Idl\cos\theta\left(\frac{\beta^3}{2\pi\omega\varepsilon_0}\right)\left(\frac{1}{(\beta r)^2} - \frac{j}{(\beta r)^3}\right)e^{-j\beta r}$$

$$E_\theta = Idl\sin\theta\left(\frac{\beta^3}{4\pi\omega\varepsilon_0}\right)\left(\frac{j}{(\beta r)} + \frac{1}{(\beta r)^2} - \frac{j}{(\beta r)^3}\right)e^{-j\beta r}$$

$$H_\phi = Idl\sin\theta\left(\frac{\beta^2}{4\pi\omega}\right)\left(\frac{j}{(\beta r)} + \frac{1}{(\beta r)^2}\right)e^{-j\beta r}$$

For the magnetic dipole,

$$H_r = IdA\cos\theta\left(\frac{\beta^3}{2\pi}\right)\left(\frac{j}{(\beta r)^2} + \frac{1}{(\beta r)^3}\right)e^{-j\beta r}$$

$$H_\theta = IdA\sin\theta\left(\frac{\beta^3}{4\pi}\right)\left(\frac{-1}{(\beta r)} + \frac{j}{(\beta r)^2} + \frac{1}{(\beta r)^3}\right)e^{-j\beta r}$$

$$E_\phi = IdA\sin\theta\left(\frac{\beta^4}{4\pi\omega\varepsilon_0}\right)\left(\frac{1}{(\beta r)} - \frac{j}{(\beta r)^2}\right)e^{-j\beta r}$$

In all the above,

β = the phase constant $2\pi/\lambda$
ω = the angular frequency of I in rad/s
ε_0 = the permittivity of free space
r, θ describe the co-ordinates of point P
E_r, E_θ, E_ϕ are the electric field vectors in V/m
H_r, H_θ, H_ϕ are the magnetic field vectors in A/m

These equations show that

a) for $\beta r << 1$ (the near field) the higher order terms dominate with E varying as $1/r^3$ and H as $1/r^2$ for the electric dipole, and vice versa for the magnetic. The $1/r^2$ terms are known as the induction field.

b) for $\beta r >> 1$ (the far field) the radial term (E_r or H_r) becomes insignificant and the transverse terms (θ and ϕ) propagate as a plane wave, varying as $1/r$.

C.4 Shielding

Skin depth

$$\delta = (\pi \cdot F \cdot \mu \cdot \sigma)^{-0.5} \text{ metres}$$

For a conductor with permeability μ_r and conductivity σ_r,

$$\delta = 0.0661 \cdot (F \cdot \mu_r \cdot \sigma_r)^{-0.5} \text{ metres}$$
$$\text{or } 2.602 \cdot (F \cdot \mu_r \cdot \sigma_r)^{-0.5} \text{ inches}$$

Properties of typical conductors [12]

Material	Relative conductivity σ_r (copper = 1)†	Relative permeability @ 1kHz * μ_r
Silver	1.08	1
Copper	1.00	1
Gold	0.70	1
Chromium	0.66	1
Aluminium	0.61	1
Zinc	0.30	1
Tin	0.15	1
Nickel	0.22	50 – 60
Mild steel	0.10	300 – 600
Mu-metal	0.03	20,000

*: relative permeability approaches 1 above 1MHz for most materials
†: absolute conductivity of copper is $5.8 \cdot 10^{-7}$ mhos

Reflection loss (R)

The magnitude of reflection loss depends on the ratio of barrier impedance to wave impedance, which in turn depends on its distance from the source and whether the field is electric or magnetic (in the near field) or whether it is a plane wave (in the far field). The following expressions are for F in Hz, r in metres and μ_r and σ_r as shown above.

$R = 168 - 10 \cdot \log_{10}((\mu_r/\sigma_r) \cdot F)$ dB .... Plane wave

$R_E = 322 - 10 \cdot \log_{10}((\mu_r/\sigma_r) \cdot F^3 \cdot r^2)$ dB .... Electric field

$R_H = 14.6 - 10 \cdot \log_{10}((\mu_r/\sigma_r)/F \cdot r^2)$ dB .... Magnetic field

Absorption loss (A)

$A = 8.69 \cdot (t/\delta)$ dB
where t is barrier thickness, δ is skin depth

Re-reflection loss (B)

$B = 20 \cdot \log_{10}(1 - e^{-2\sqrt{2}(t/\delta)})$ dB

B is negligible unless material thickness t is less than the skin depth δ;
e.g. if $t = \delta$, $B = -0.53$dB; if $t = 2\delta$, $B = -0.03$dB
B is always a negative value, since multiple reflections degrade shielding effectiveness

Shielding effectiveness

$SE_{dB} = R_{dB} + A_{dB} + B_{dB}$

C.5 Capacitance, inductance and pcb layout

Capacitance

Capacitance between two plates of area A cm^2 spaced d cm apart in free space

$C = 0.0885 \cdot A/d$ pF

The self-capacitance of a sphere of radius r cm

$C = 4\pi \cdot 0.0885 \cdot r = 1.1 \cdot r$ pF

The capacitance per unit length between concentric circular cylinders of inner radius r_1, outer radius r_2 in free space

$C = 2\pi \cdot 0.0885 / \ln(r_2/r_1)$ pF/cm

The capacitance per unit length between two conductors of diameter d spaced D apart in free space

$C = \pi \cdot 0.0885 / \cosh^{-1}(D/d)$ pF/cm

The factor 0.0885 in each of the above equations is due to the permittivity of free space ε_0; multiply by the dielectric constant or relative permittivity ε_r for other materials

Inductance [5]

The inductance of a straight length of wire of length l and diameter d

$L = 0.0051 \cdot l \cdot (\ln(4l/d) - 0.75)$ μH for l, d in inches, or
$0.002 \cdot l \cdot (\ln(4l/d) - 0.75)$ μH for l, d in cm

A useful rule of thumb is 20nH/inch.

The inductance of a return circuit of parallel round conductors of length l cm, diameter d and distance apart D, for $D/l \ll 1$

$$L = 0.004 \cdot l \cdot (\ln(2D/d) - D/l + 0.25) \ \mu H$$

The mutual inductance between two parallel straight wires of length l cm and distance apart D, for $D/l \ll 1$

$$M = 0.002 \cdot l \cdot (\ln(2l/D) - 1 + D/l) \ \mu H$$

The mutual inductance between two conductors spaced D apart at height h over a ground plane carrying its return current [59]

$$M = 0.001 \cdot \ln(1 + (2h/D)^2) \ \mu H/cm$$

The inductance of a single wire of diameter d at height h over a ground plane carrying its return current [59]

$$L = 0.002 \cdot \ln(4h/d) \ \mu H/cm$$

PCB track propagation delay and characteristic impedance [60]

Surface microstrip

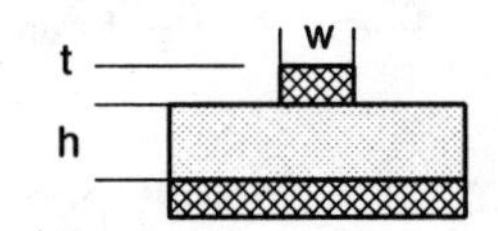

$$T_{pd} = 1.017 \cdot \sqrt{(0.475 \cdot \varepsilon_r + 0.67)} \ ns/ft$$

$$Z_0 = (87/\sqrt{(\varepsilon_r + 1.41)}) \cdot \ln[5.98h/(0.8w + t)] \ \Omega$$

For h = 1.6mm, w = 0.3mm and t << w in surface microstrip, $Z_0 = 132\Omega$

Embedded stripline

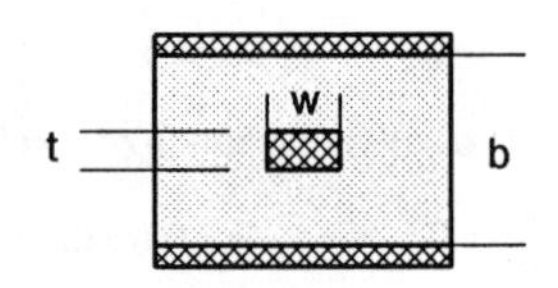

$$T_{pd} = 1.017 \cdot \sqrt{\varepsilon_r} \ ns/ft$$

$$Z_0 = (60/\sqrt{\varepsilon_r}) \cdot \ln[4b/(0.67\pi \cdot (0.8w + t))] \ \Omega$$

For FR4 epoxy fibreglass pcb material ε_r is typically 4.5 which gives a propagation delay T_{pd} of 1.7ns/ft for surface microstrip and 2.15ns/ft for embedded stripline

For b = 1.4mm, w = 0.3mm and t << w in embedded stripline, Z_0 is 68Ω

When a track is loaded with devices, their capacitances modify the track's propagation delay and Z_0 as follows:

$$T_{pd}' = T_{pd} \cdot \sqrt{(1 + C_D/C_o)}$$

$$Z_0' = Z_0/\sqrt{(1 + C_D/C_o)}$$

where C_D is the distributed device capacitance per unit length, i.e. the total load capacitance divided by the track length, and C_o is the intrinsic capacitance of the track calculated from

$$C_o = 1000 \cdot (T_{pd}/Z_0) \ pF/length$$

C.6 Filters

Second order low pass filter [76]

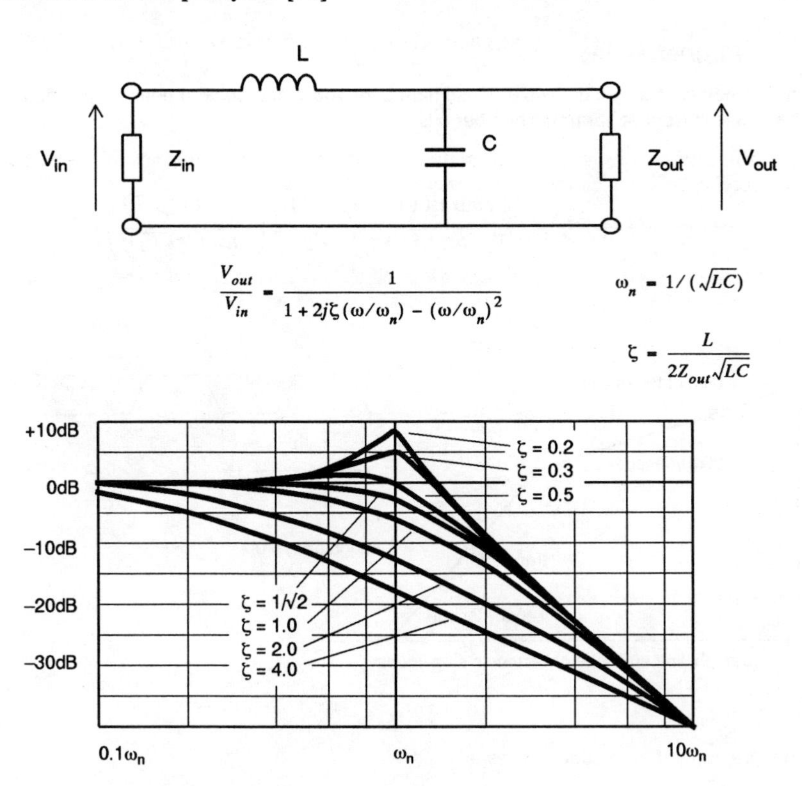

The damping factor ζ describes both the insertion loss at the corner frequency and the frequency response of the filter. ζ is affected by the load impedance and low values may cause insertion gain around the corner frequency.

The following design procedure may be applied to any low pass LC filter and to the typical mains filter configuration (see Figure 6.22 on page 190) to design both the differential components and the common mode components, remembering that the latter are symmetrical about earth and can therefore be treated as two separate circuits.

1. Identify the required cutoff (corner) frequency ω_n: the second order filter rolls off at 40dB/decade, so the desired attenuation A_{dB} at some higher frequency F will put the corner frequency at

   $\omega_n = 2\pi F / \log^{-1}(A/40)$

2. Identify the load resistance Z_{out} and desired damping factor ζ. A value for ζ between 0.7 and 1 will normally be adequate if Z_{out} is reasonably well specified. Values much larger than 1 will cause excessive low frequency attenuation while much less than 0.7 will cause ringing and insertion gain.

3. From these calculate the required component values:

   $L = 2 \cdot Z_{out} \cdot \zeta / \omega_n$

$C = 1/(L \cdot \omega_n^2)$

4. Iterate as required to obtain useable standard component values.

C.7 Fourier series

For a symmetrical trapezoidal wave of rise time t_r, period T and peak to peak amplitude I, the harmonic current at harmonic number n is

$$I(n) = 2I((t+t_r)/T)\left(\frac{\sin n\pi((t+t_r)/T)}{n\pi((t+t_r)/T)}\right)\left(\frac{\sin n\pi(t_r/T)}{n\pi(t_r/T)}\right)$$

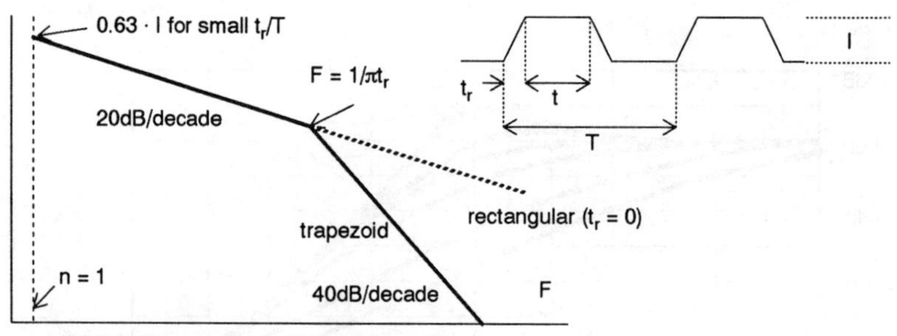

Straight line envelope of harmonic amplitudes

The general form of the Fourier Series is

$$f(t) = 0.5A_0 + \sum_{n=1}^{\infty}(A_n \cos\omega_n t + B_n \sin\omega_n t)$$

where the coefficients A_n and B_n are

$$A_n = \frac{2}{T}\int_{-T/2}^{T/2} f(t)\cos\omega_n t\,dt$$

$$B_n = \frac{2}{T}\int_{-T/2}^{T/2} f(t)\sin\omega_n t\,dt$$

Any arbitrary waveform can be analysed by sampling it at discrete time intervals and taking the discrete Fourier transform (DFT). This is achieved by replacing the integrals above by a finite weighted summation:

$$A(m) = \frac{1}{N}\sum_{n=0}^{N-1} x(n)\cos 2\pi m\left(\frac{n}{N}\right)$$

$$B(m) = \frac{1}{N}\sum_{n=0}^{N-1} x(n)\sin 2\pi m\left(\frac{n}{N}\right)$$

Here the time axis is given by (n/N) where N is the total number of samples and x(n) is the sample value at the nth sample. m represents the frequency axis and the DFT calculates A(m) and B(m) for each discrete frequency from m = 0 (DC) through to m = (N/2) – 1.

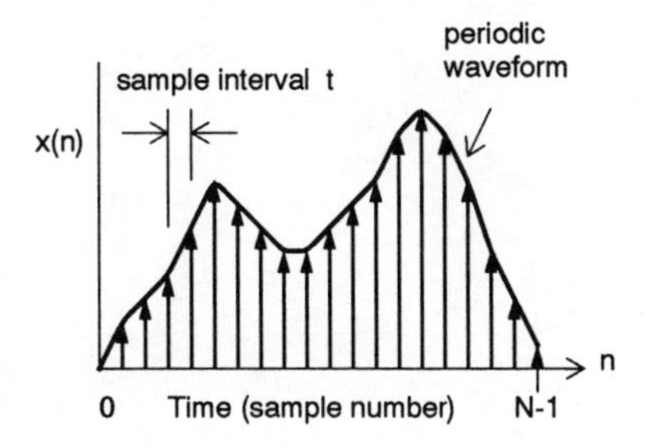

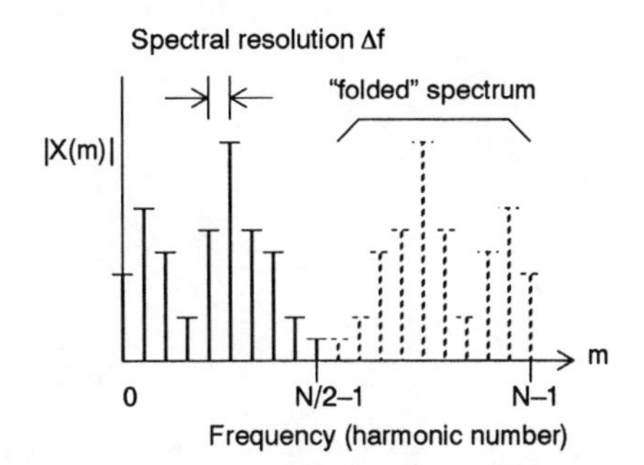

The spectral resolution Δf in the frequency domain is the reciprocal of the total sample time, 1/Nt, which is equivalent to the period of the time domain waveform. m = 1 represents the fundamental frequency, m = 2 the second harmonic and so forth. The spectrum image is folded about m = N/2 and therefore the maximum harmonic frequency that can be analysed is half the number of samples times the fundamental, or 1/2t.

For EMC work A(0) and B(0) are normally neglected (they represent the DC component) as also is phase information, represented by the phase angle between the real and imaginary components arg(A(m) + jB(m)). The mean square amplitude

$$|X(m)|^2 = |A(m)|^2 + |B(m)|^2$$

represents the power in the mth harmonic. A simple program to calculate X(m) for an array of samples x(n) in the time domain is easily written [47].

Appendix D

CAD for EMC

D.1 Overview

It may seem strange to devote no more than a few pages at the back of a book on product design for EMC to the important subject of computer aided design (CAD). The prospect of non-compliance with EMC requirements is now sufficiently threatening that manufacturers are having to give these requirements serious consideration at the design stage, and the second part of this book discusses the major design principles that this involves. Many other aspects of the circuit design process are now automated and simulated to the extent that a breadboard stage, to check the correctness of the basic design concepts, is often no longer necessary. A very attractive option to the product designer would be a CAD tool that predicted RF emissions and susceptibility, with enough accuracy for initial evaluation purposes, from the design data of the product. This would allow alternative EMC techniques to be tried out before the costly stage of committing to tooling and pre-production had begun.

The reason why this subject is relegated to an appendix is because no such tool yet exists. Many problems of a specific nature can be solved by electromagnetics computation packages that have been available for many years, but these do not address the needs of product designers. On the other hand, several groups are working on the production of software packages that are tailored to these needs, and the next few years may well see the successful introduction of such tools.

The difficulties facing these researchers are many:

- modelling a small collection of electromagnetic emitting elements (usually a current segment) is easy. As the number of elements n increases, the computational memory increases as n^2 and the processing time as n^3, so that computer speed and storage limitations place major restrictions on the size of the problem being handled.
- as the frequency increases, so the element size must be reduced to maintain accuracy, further increasing the required number of elements. Typical computational methods work in the frequency domain and so the calculations must be repeated for each frequency of interest.
- PCBs with ground plane layers can effectively be approximated as conducting sheets, which simplifies the structure model, whereas those without ground planes cannot be so easily modelled.
- connecting wires and cables have a large effect on the coupling and must be modelled explicitly. This may invalidate the results if the layout is not properly defined in the final product. Some early attempts modelled the performance of the pcb only and their results bear little relevance to reality.
- the driving currents can theoretically be derived from device models and

calculated track impedances, but these may differ substantially from the real circuit with real tolerances. For example the amplitudes of higher order clock harmonics depend heavily on risetime, which is a poorly specified parameter and is affected by circuit capacitances and device spread.

These difficulties relate only to emission predictions. A further set of variables come into play when you try to simulate susceptibility, and apart from some research into immunity at the IC level these have hardly been addressed yet. There are other obstacles to implementing practical tools, which are not related to the technical problems of electromagnetic modelling [57]:

- EMC design aspects do not respect the demarcation common in design labs of circuit, layout and mechanical disciplines. Considerable interaction is needed between these which presupposes a common body of EMC knowledge among the different fields, which is unlikely to exist.
- To overcome this demarcation, the CAD package must integrate these aspects and must therefore accept input on all fronts. Few integrated CAD environments are installed which could provide such input automatically, but manual input is not realistic given the time constraints facing a typical design department.
- The output of the package must also be in a form which is useable and comprehensible by these different designers. It must be structured to be of maximum assistance at the most appropriate phase(s) of the design process. Some re-training of the users may be required so that they can actually use the output.

D.2 Available packages

Those software packages which are currently available for electromagnetic modelling purposes have been developed for solving EMI coupling problems in certain well-defined applications. Every EMC problem can be described in terms of a source of interference, a coupling path and a receptor (or victim) of the interference. The structure of the coupling path may include either or both radiative or conductive mechanisms, and these are often amenable to analysis so that, for instance, the voltage or current present at an interface with an item of equipment can be derived as a result of coupling of the structure containing that equipment with an external field.

A typical application may be to model the surface currents on an aircraft fuselage which is illuminated by a plane-wave field and deduce the currents which may flow in the cable bundles within the fuselage. Another approach is to determine the RF energy transport by penetrations through a shield such as electrical conductors or pipework. The codes which perform the electromagnetic modelling of these situations use one or a combination of finite difference, method of moments, transmission line modelling or finite element methods to solve Maxwell's equations directly. Each has its advantage in a particular situation, for instance the method of moments code NEC deals efficiently with wire coupling problems such as are common in antenna design, whereas the finite element code EMAS or transmission line modelling (TLM) are more suited to problems which have inhomogeneous regions and complex geometries, such as surface currents on enclosures.

Many approximations need to be made even for well-defined problems to allow computation with a reasonable amount of effort. For instance near field coupling is much more complicated than far field, since the nature of the source strongly affects the incident wave. Shield aperture size is critical since for electrically large apertures (size comparable to a wavelength) the internal and external regions must be treated consistently, whereas the modelling of fields penetrating through small apertures can be simplified.

The major difficulty with applying these packages to commercial product design is that they have been developed for a rather different set of purposes. Mostly they have been derived from research on military EMC problems. This has two consequences:

- technically, they are appropriate only for situations which are clearly defined: cable routes, connector terminations, mechanical structures and shielded enclosures are all carefully designed and controlled through to the final product and its installation. This contrasts with the commercial environment where the costs of doing this, and indeed the impossibility in many cases, puts such an approach out of court. Thus the parameters and approximations that would have to be made in the simulation are not valid in reality.
- operationally, there are difficulties in performing the simulation. Few of the packages (which started life in university research departments) have been developed to the point of being user-friendly. A great deal of input data, usually in the form of structure co-ordinates, must be loaded and validated before the program can run, and its output must then be interpreted. A typical product design engineer will have neither the time nor the training to do this – only those companies with the resources to run a specialist group devoted to the task can handle it successfully.

Nevertheless, these packages can find a use in areas of EMC not directly related to product design, such as those to do with radiated field testing: predicting reflections in screened rooms and damping them, predicting proximity effects between the antenna and the EUT, and predicting calibration errors when antennas are used in screened rooms [30].

D.3 Circuit CAD

Some circuit design and PCB layout CAD packages already offer transmission line analysis for high speed logic design, taking into account track parameters calculated from their geometries and the board dielectric. Extending these capabilities into the domain of RF properties of the board interacting with its environment – as will be needed for EMC prediction – will be difficult. The author knows of two efforts which are being made in this direction at the time of writing, and there are no doubt more:

- the EMC Workbench project under the auspices of JESSI (project AC5) [58] which has several major European industrial partners; its models will include radiation from 3D structures, and it was started in 1991;
- a project sponsored by Lynwood Scientific Developments and the UK DTI, being carried out at the University of York [57], which is aimed at predicting PCB emissions in a realistic environment, i.e. including the presence of cables and conducting bodies, and housed in a poorly shielded or unshielded enclosure.

Appendix E

The EC and EFTA countries

E.1 The European Community

Established by the Treaty of Rome, 1957. * = founder members

Belgium	*
Germany	*
Denmark	
Spain	
France	*
Greece	
Ireland	
Italy	*
Luxembourg	*
The Netherlands	*
Portugal	
United Kingdom	

Applications for membership have been received from Austria, Cyprus, Malta, Sweden and Turkey

E.2 The European Free Trade Association

Austria
Norway
Sweden
Switzerland
Finland
Iceland
Liechtenstein

Formal negotiations on the establishment of a European Economic Area (EEA) encompassing all 19 EC and EFTA countries began in 1990. These negotiations were scheduled to be completed by the end of 1991 so that the EEA treaty can come into force at the same time as the Single European Market in 1993.

Glossary

AM	Amplitude modulation
ASIC	Application Specific Integrated Circuit
BBC	British Broadcasting Corporation
BCI	Bulk current injection
CAD	Computer aided design
CB	Citizen's band
CCTV	Closed circuit television
CEN	European Organisation for Standardization
CENELEC	European Organisation for Electrotechnical Standardization
CISPR	International Special Committee for Radio Interference
CMR	Common mode rejection
DTI	Department of Trade and Industry
EC	European Commission
EED	Electro-explosive device
EEPROM	Electrically erasable programmable read-only memory
EFTA	European Free Trade Association
EMC	Electromagnetic compatibility
EMI	Electromagnetic interference
EMR	Electromagnetic radiation
EOTC	European Organization for Testing and Certification
ESD	Electrostatic discharge
ETSI	European Telecommunications Standards Institute
EUT	Equipment under test
FCC	Federal Communications Commission
HF	High frequency
I/O	Input/output
IEC	International Electrotechnical Commission
IEEE-488	IEEE standard for data communications between test instruments
IF	Intermediate frequency
ISDN	Integrated Services Digital Network
ISM	Industrial, scientific and medical
ITE	Information technology equipment
ITU	International Telecommunications Union
LF	Low frequency
LISN	Line impedance stabilising network
MF	Medium frequency
MS	Mains signalling
NAMAS	National Measurement Accreditation Service
NRPB	National Radiological Protection Board
OATS	Open area test site
OJEC	Official Journal of the European Communities

PC	Personal computer
PCB	Printed circuit board
PFC	Power factor correction
PRF	Pulse repetition frequency
RAM	Random access memory (also RF absorbent material)
RF	Radio frequency
RFI	Radio frequency interference
RMS	Root mean square
ROM	Read only memory
SI	Statutory Instrument (also Système Internationale)
SMPS	Switched mode power supply
SMT	Surface mount technology
STI	Surface transfer impedance
TEM	Transverse electromagnetic
TTE	Telecommunications terminal equipment
UART	Universal asynchronous receiver/transmitter
VDE	Verband Deutscher Elektrotechniker (Association of German Electrical Engineers)
VDR	Voltage dependent resistor
VDU	Visual display unit
VLSI	Very large scale integration
VSWR	Voltage standing wave ratio
ZVEI	Zentralverband der Elektrotechnischen Industrie (Germany)

Bibliography

Books

[1] **BBC Radio Transmitting Stations 1991; BBC Television Transmitting Stations 1990**
BBC Engineering Information, White City, London

[2] **A Handbook of Fourier Theorems**
D C Champeney, Cambridge University Press, 1987

[3] **An Introduction to Applied Electromagnetism**
Christos Christopoulos, University of Nottingham, Wiley, 1990

[4] **Discrete Fourier Transforms and their Applications**
V Cízek, Adam Hilger, 1986

[5] **Inductance Calculations**
F W Grover, Van Nostrand, New York 1946

[6] **Engineering Electromagnetics**
William H Hayt, 5th edition, McGraw Hill, 1988

[7] **Electronic Engineer's Reference Book**
F F Mazda (ed), 5th edn, Butterworth, 1983

[8] **Grounding and Shielding Techniques in Instrumentation**
Ralph Morrison, 3rd Edition, Wiley, 1986

[9] **FACT Advanced CMOS Logic Databook**
National Semiconductor, 1990

[10] **Noise Reduction Techniques in Electronic Systems**
Henry W Ott, Bell Labs, 2nd edition, Wiley, 1988

[11] **Applied Electromagnetics**
Martin A Plonus, McGraw Hill, 1978

[12] **Design Guide to the Selection and Application of EMI Shielding Materials**
TECKNIT EMI Shielding Products, 1991

[13] **Supplement to TTL Data Book Vol 2: Advanced Schottky**
Texas Instruments, 1984

[14] **Mathematical Methods in Electrical Engineering**
Thomas B A Senior, Cambridge University Press, 1986

[15] **Coupling of External Electromagnetic Fields to Transmission Lines**
A A Smith, IBM, Wiley, 1977

[16] **Antenna Theory and Design**
W L Stutzman, G A Thiele, Wiley, 1981

[17] **Electromagnetic Shielding Materials and Performance**
D R J White, Don White Consultants Inc, 2nd Edition, 1980

[18] **The Circuit Designer's Companion**
T Williams, Butterworth Heinemann, 1991

Papers

[19] **EMC antenna calibration and the design of an open space antenna range**
M J Alexander, NPL, 1989 British Electromagnetic Measurements Conference, NPL, Teeddington, 7-9 November 1989

[20] **A Novel Technique for Damping Site Attenuation Resonances in Shielding Semi-Anechoic Rooms**
B H Bakker, H F Pues, Grace NV, IEE 7th International Conference on EMC, York 28-31st Aug 1990 pp119-124

[21] **Improving the mains transient immunity of microprocessor equipment**
S J Barlow, Cambridge Consultants, IEE Colloquium, Interference And Design for EMC in Microprocessor Based Systems, London 1990

[22] **Implementation of the EMC Directive in the UK**
A E J Bond, DTI, EMC 91 – Direct to Compliance, ERA Technology Conference, Heathrow, February 1991

[23] **Coated Plastics Offer Shielding and Savings**
R Brander, P Kuzyk, R Bellemare, Enthone-OME, EMC Technology, July/August 1990, pp37-40

[24] **Scan Speed Limits in Automated EMI Measurements**
E L Bronaugh, Electro-Metrics, IERE 5th International Conference on EMC, York, 1986

[25] **A Simple Way of Evaluating the Shielding Effectiveness of Small Enclosures**
D R Bush, IBM, 8th Symposium on EMC, Zurich, March 5-7 1989

[26] **Everybody needs standards**
I Campbell, BSI, IEE News 3rd October 1991 p14

[27] **Designing for EMC with HCMOS Microcontrollers**
M Catherwood, Motorola Application Note AN1050, 1989

[28] **Coupling Reduction in Twisted Wire**
W T Cathey, R M Keith, Martin Marietta, International Symposium on EMC, IEEE, Boulder, Aug 1981

[29] **A Review of Transients and their Means of Suppression**
S Cherniak, Motorola Application Note AN-843, 1982

[30] **Numerical Modelling in EMC**
C Christopoulos (guest editor), University of Nottingham, International Journal of Numerical Modelling (special issue), Vol 4 No 3, September 1991

[31] **Unscrambling the mysteries about twisted wire**

R B Cowdell, International Symposium on EMC, IEEE, San Diego, October 9-11 1979

[32] **Common Mode Filter Inductor Analysis**

L F Crane, S F Srebranig, Coilcraft, in Coilcraft Data Bulletin, 1985

[33] **Alternative Methods of Damping Resonances in a Screened Room in the Frequency Range 30 to 200MHz**

L Dawson, A C Marvin, University of York, IEE 6th International Conference on EMC, York, 1988

[34] **A Cost Effective Path to Electromagnetic Compatibility**

S J Ettles, British Telecom, EMC 90 – The Achievement of Compatibility, ERA Technology Conference Proceedings 90-0089, July 1990

[35] **Cúk: the best SMPS?**

T S Finnegan, Electronics & Wireless World, January 1991, pp 69 – 72

[36] **Systems Installations EMC Considerations**

A Finney, GPT, Audio Engineering Society Conference, Will You be Legal?, London, 19th March 1991

[37] **The GTEM Cell concept: applications of this new EMC test environment to radiated emission and susceptibility measurements**

H Garbe, D Hansen, ABB, IEE 7th International Conference on EMC, York 28-31st Aug 1990 pp 152-156

[38] **Use of a Ground Grid to reduce Printed Circuit Board Radiation**

R F German, IBM, 6th Symposium on EMC, Zurich, March 5-7 1985

[39] **Design Innovations Address Advanced CMOS Logic Noise Considerations**

M J Gilbert, National Semiconductor Application Note AN-690, 1990

[40] **Transients in Low-Voltage Supply Networks**

J J Goedbloed, Philips, IEEE Transactions on Electromagnetic Compatibility, Vol EMC-29, No 2 May 1987, pp104 - 115

[41] **Characterization of Transient and CW Disturbances induced in Telephone Subscriber Lines**

J J Goedbloed, W A Pasmooij, Philips Research, IEE 7th International Conference on EMC, York 28-31st Aug 1990 pp211-218

[42] **A Simple RF Immunity Test Set-up**

P Groenveld, A de Jong, Philips Research, IEEE 2nd Symposium on EMC, Montreux, June 28-30 1977, pp 233-239

[43] **A Broadband Alternative EMC Test Chamber Based on a TEM-Cell Anechoic Chamber Hybrid Concept**

D Hansen et al, ABB Ltd, IEEE 1989 International Symposium on EMC, Nagoya, Japan, 8-10 September 1989 pp 133-137

[44] **EMC related to the public electricity supply network**

G O Hensman, Electricity Council, EMC 89 - Product Design for Electromagnetic Compatibility, ERA Technology Seminar Proceedings 89-0001

[45] **Techniques for Supply Harmonic Measurement**

G P Hicks, ERA Technology, IEE Colloquium "Single-Phase Supplies: Harmonic Regulation and Remedies", 27th March 1991, IEE Colloquium Digest no 1991/071

[46] **How Switches Produce Electrical Noise**

E K Howell, General Electric, IEEE Transactions on Electromagnetic Compatibility, Vol EMC-21, No 3 August 1979, pp162 – 170

[47] **Interfacing with C (part 8)**

H Hutchings, Electronics & Wireless World, December 1990, pp1066 – 1072

[48] **Survey of EMC measurement techniques**

G A Jackson, ERA Technology, Electronics & Communication Engineering Journal, March/April 1989, pp 61 - 70

[49] **On the Design and Use of the Average Detector in Interference Measurement**

W Jennings, H Kwan, DTI, 5th International Conference on EMC, York, 1986, IERE Conf Pub 71

[50] **Achieving Compatibility in Inter-Unit Wiring**

J W E Jones, Portsmouth Poly, 6th Symposium on EMC, Zurich, March 5-7 1985

[51] **Co-ordination of IEC standards on EMC and the importance of participating in standards work**

R Kay, IEC, IEE 7th International Conference on EMC, York 28-31st Aug 1990, pp1-6

[52] **Legal Aspects of the EMC Directive**

J F C Ketchell, DTI, EMC 89 - Product Design for Electromagnetic Compatibility, ERA Technology seminar proceedings 89-0001, 7th February 1989

[53] **Average Measurements using a Spectrum Analyzer**

S Linkwitz, Hewlett Packard, EMC Test & Design, May/June 1991 pp 34-38

[54] **The Antenna Cable as a Source of Error in EMI Measurements**

J DeMarinis, DEC, International Symposium on EMC, IEEE, Washington, Aug 2-4 1988

[55] **Getting Better Results from an Open Area Test Site**

J DeMarinis, DEC, 8th Symposium on EMC, Zurich, March 1989

[56] **Convenient Current-Injection Immunity Testing**

R C Marshall, Protech, IEE 7th International Conference on EMC, York 28-31st Aug 1990, pp173-176

[57] **Computer aided design for radio frequency interference reduction**

A C Marvin et al, University of York, IEE 7th International Conference on EMC, York 28-31st Aug 1990

[58] **FACET – Simulation of EMC effects in PCBs and MMICs**

R F Milsom, Philips, IEE discussion meeting on The Application of Field Computation Packages to Industrial EMC Problems, London, 15th October 1991

[59] **Coupling between Open and Shielded Wire Lines over a Ground Plane**

R J Mohr, IEEE Transactions on Electromagnetic Compatibility, Vol EMC-9 No 2, September 1967, Appendix 1

[60] **Transmission Line Effects in PCB Applications**

Motorola, Application Note AN1051, 1990

[61] **Car makers seek new ICs to improve EMC**
New Electronics, November 1991, p.14

[62] **A Structured Design Methodology for Control of EMI Characteristics**
U Nilsson, EMC Services AB, IEE 7th International Conference on EMC, York 28-31st Aug 1990

[63] **Ground - A Path for Current Flow**
H W Ott, Bell Labs, International Symposium on EMC, IEEE, San Diego, October 9-11 1979

[64] **Digital Circuit Grounding and Interconnection**
H W Ott, Bell Labs, International Symposium on EMC, IEEE, Boulder, Aug 1981

[65] **Controlling EMI by Proper Printed Wiring Board Layout**
H W Ott, Bell Labs, 6th Symposium on EMC, Zurich, March 5-7 1985

[66] **Shielded Flat Cables for EMI and ESD Reduction**
C M Palmgren, 3M, International Symposium on EMC, IEEE, Boulder, Aug 1981

[67] **High Frequency Series Resonant Power Supply - Design Review**
R Patel, R Adair, Unitrode ST-A2, in Unitrode Applications Handbook, 1985

[68] **Effect of Pigtails on Crosstalk to Braided-Shield Cables**
C R Paul, Univ of Kentucky, IEEE Transactions on Electromagnetic Compatibility, Vol EMC-22 No 3, Aug 80

[69] **Diagnosis and Reduction of Conducted Noise Emissions**
C R Paul, K B Hardin, IBM, International Symposium on EMC, IEEE, Washington, Aug 2-4 1988

[70] **A Comparison of the Contributions of Common-Mode and Differential-Mode Currents in Radiated Emissions**
C R Paul, Univ of Kentucky, IEEE Transactions on Electromagnetic Compatibility, Vol EMC-31 No2 May 89

[71] **The approach to the implementation of the EMC Directive in Germany – problems and perspectives**
D Rahmes, BPT, EMC 91 – Direct to Compliance, ERA Technology conference, London, February 1991

[72] **Screening Plastics Enclosures**
I Rankin, PERA, Suppression Components, Filters and Screening for EMC, ERA Technology Seminar Proceedings 86-0006

[73] **Radiation Susceptibility of Digital Commercial Equipment**
W T Rhoades, Xerox, International Symposium on EMC, IEEE, Washington, Aug 1983

[74] **EMC Measurement Uncertainties**
M F Robinson, British Telecom, 1989 British Electromagnetic Measurements Conference, NPL, Teeddington, 7-9 November 1989

[75] **Standard Site Method for Determining Antenna Factors**
A A Smith, IBM, IEEE Transactions on Electromagnetic Compatibility, Vol EMC-24 No 3, Aug 82

[76] **Guide for Common Mode Filter Design**
S F Srebranig, L F Crane, Coilcraft, in Coilcraft Data Bulletin, 1985

[77] **Designing for Electrostatic Discharge Immunity**
D M Staggs, EMCO, 8th Symposium on EMC, Zurich, March 1989

[78] **The Generic Standards**
R Storrs, Televerket, EMC 91 – Direct to Compliance, ERA Technology conference, London, February 1991

[79] **Demodulation RFI in inverting and non-inverting operational amplifier circuits**
Y-H Sutu, J J Whalen, SUNY, 6th Symposium on EMC, Zurich, March 1985

[80] **Radiated Emission, Susceptibility and Crosstalk Control on Ground Plane Printed Circuit Boards**
A J G Swainson, Thorn EMI Electronics, IEE 7th International Conference on EMC, York 28-31st Aug 1990 pp37-41

[81] **RFI Susceptibility evaluation of VLSI logic circuits**
J G Tront, 9th Symposium on EMC, Zurich, March 1991

[82] **Design Philosophy for Grounding**
P C T Van Der Laan, M A Van Houten, 5th International Conference on EMC, York, 1986, IERE Conf Pub 71

[83] **The Availability of Standards**
M C Vrolijk, Nederlandse Philips Bedrijven BV, EMC 91 - Direct to Compliance, ERA Technology, February 1991

[84] **Determining EMI in Microelectronics - A review of the past decade**
J J Whalen, SUNY, 6th Symposium on EMC, Zurich, March 5-7 1985

[85] **The development of EMI design guidelines for switched mode power supplies - examples and case studies**
M Wimmer, Schrack Elektronik AG, 5th International Conference on EMC, York, 1986, IERE Conf Pub 71

Official publications and standards

European standards: available in the UK from

British Standards Institution
Sales Department
Linford Wood
Milton Keynes
MK14 6LE
UK
Tel (0908) 221166; Telex 825777 BSIMK G; Telefax 0908 320856

[86] **EN55011: Limits and methods of measurement of radio disturbance characteristics of industrial, scientific and medical (ISM) radio-frequency equipment**

[87] **EN55013: Limits and methods of measurement of radio disturbance characteristics of broadcast receivers and associated equipment**

[88] **EN55014: Limits and methods of measurement of radio interference characteristics of household electrical appliances, portable tools and similar electrical apparatus**

[89] **EN55015: Limits and methods of measurement of radio interference characteristics of fluorescent lamps and luminaires**

[90] **EN55020: Immunity from radio interference of broadcast receivers and associated equipment**

[91] **EN55022: Limits and methods of measurement of radio interference characteristics of information technology equipment**

[92] **EN60555: Disturbances in supply systems caused by household appliances and similar electrical equipment**

[93] **EN50081: Electromagnetic compatibility - generic emission standard**

[94] **EN50082: Electromagnetic compatibility - generic immunity standard**

[95] **pr EN55101: Immunity requirements for Information Technology Equipment**

[96] **pr ETS 300127: Equipment engineering: Radiated emission testing of physically large systems**

[97] **Draft: EMC tests in shielded anechoic enclosures part 1: requirements for shielded anechoic enclosures for EMC tests**
CLC/TC110(SEC)61, BSI document no 91/30259DC

IEC standards: available from BSI as above or direct from

IEC Central Office
1 Rue de Varembé
1211 Geneva 20
Switzerland
Tel (022) 340150; Telex 28872 CEIEC-CH; Telefax (022) 333843

[98] **IEC50 (161): (BS4727 : Pt 1 : Group 09) Glossary of electrotechnical, power, telecommunication, electronics, lighting and colour terms: Electromagnetic compatibility**

[99] **IEC801: (BS6667) Electromagnetic compatibility for industrial-process measurement and control equipment**

[100] **IEC1000: Electromagnetic compatibility**

[101] **CISPR##: see EN550## above**

[102] **CISPR16: (BS727) Specification for radio interference measuring apparatus**

[103] **CISPR23: Determination of limits for industrial, scientific and medical equipment**

Other British Standards

[104] **BS1597: Limits and methods of measurement of electromagnetic interference generated by marine equipment and installations**

[105] **BS5049: (CISPR18-2) Methods for measurement of radio interference characteristics of overhead power lines and high voltage equipment**

[106] **BS5602: (CISPR18-1, -3) Code of practice for abatement of radio interference from overhead power lines**

[107] **BS6656: Guide to prevention of inadvertent ignition of flammable atmospheres by radio-frequency radiation**

[108] **BS6657: Guide for prevention of inadvertent initiation of electro-explosive devices by radio-frequency radiation**

[109] **BS6839: Part 1: (EN50065-1) Mains signalling equipment: specification for communication and interference limits and measurements**

[110] **BS7027: Limits and methods of measurement of immunity of marine electrical and electronic equipment to conducted and radiated electromagnetic interference**

[111] **Aero 3G100 : Part 4 : Section 2 : General requirements for equipment for use in aircraft: electromagnetic interference at radio and audio frequencies**

The EMC Directive

[112] **Council Directive of 3rd May 1989 on the approximation of the laws of the Member States relating to Electromagnetic Compatibility (89/336/EEC)**
Official Journal of the European Communities No L 139, 23rd May 1989

[113] **Electrical Interference: A Consultative Document (The Implementation in the United Kingdom of Directive 89/336/EEC on Electromagnetic Compatibility)**
Department of Trade & Industry, Radiocommunications Division, 1989

[114] **The single market: electromagnetic compatibility**
Department of Trade & Industry, July 1989

[115] **The single market: the "new approach" to technical harmonisation and standards**
Department of Trade & Industry, June 1989

[116] **Electromagnetic Compatibility: latest developments**
Department of Trade & Industry, August 1991

[117] **Electromagnetic Compatibility: European Commission explanatory document on Council Directive 89/336/EEC**
Department of Trade & Industry, November 1991

[118] **Council Decision of 13th December 1990 concerning the modules for the various phases of the conformity assessment procedures which are intended to be used in the technical harmonisation directives (90/683/EEC)**
Official Journal of the European Communities No L 380/13, 31st December 1990

[119] **CE Mark: Proposed Council Regulation: consultative document**
Department of Trade & Industry, July 1991

[120] **Commission Communication in the Framework of the Implementation of the 'New Approach' Directives: Publication of titles and references of European harmonized standards complying with the essential requirements**
Official Journal of the European Communities No C 44/12, 19th February 1992

Miscellaneous official publications

[121] **Uncertainties of Measurement for NAMAS Electrical Product Testing Laboratories**
NAMAS Information Sheet NIS 20, NAMAS Executive, NPL, Teddington, July 1990

[122] **Guidance as to Restrictions on Exposures to Time Varying Electromagnetic Fields and the 1988 Recommendations of the International Non-Ionizing Radiation Committee**
NRPB-GS11, National Radiological Protection Board, Chilton, May 1989

[123] **Guidelines on limits of exposure to radiofrequency electromagnetic fields in the frequency range 100kHz to 300GHz**

INIRC, Health Physics Vol. 54, No 1 (January), pp 115-123, 1988

[124] **The UK Market for EMC Testing and Consultancy Services**

W S Atkins Management Consultants, Department of Trade & Industry, April 1989

US publications, available from:

International Transcription Services
1919 M Street, N.W.
Room 248
Washington, D.C 20554
Tel: (202) 296 7322

[125] **Understanding the FCC Regulations Concerning Computing Devices**

OST Bulletin No 62

[126] **FCC Procedure for Measuring RF Emissions from Computing Devices**

FCC/OET MP-4

[127] **Characteristics of Open Field Test Sites**

FCC Bulletin OST 55

Index